Lecture Notes in Artificial Intelligence 1812

Subseries of Lecture Notes in Computer Science
Edited by J. G. Carbonell and J. Siekmann

Lecture Notes in Computer Science
Edited by G. Goos, J. Hartmanis and J. van Leeuwen

T0238477

Springer

Berlin
Heidelberg
New York
Barcelona
Hong Kong
London
Milan
Paris
Singapore
Tokyo

Jeremy Wyatt   John Demiris   (Eds.)

Advances in Robot Learning

8th European Workshop on Learning Robots, EWLR-8
Lausanne, Switzerland, September 18, 1999
Proceedings

Springer

Series Editors

Jaime G. Carbonell, Carnegie Mellon University, Pittsburgh, PA, USA
Jörg Siekmann, University of Saarland, Saarbrücken, Germany

Volume Editors

Jeremy Wyatt
University of Birmingham
School of Computer Science
Birmingham B15 2TT, UK
E-mail: jlw@cs.bham.ac.uk

John Demiris
University of Edinburgh
Division of Informatics
Institute of Perception, Action and Behaviour
Scotland, UK
E-mail: johnde@dai.ed.ac.uk

Cataloging-in-Publication Data applied for

Die Deutsche Bibliothek - CIP-Einheitsaufnahme

Advances in robot learning : proceedings / 8th European Workshop on
Learning Robots, EWLR-8, Lausanne, Switzerland, September 18, 1999.
Jeremy Wyatt ; John Demiris (ed.). - Berlin ; Heidelberg ; New York ;
Barcelona ; Hong Kong ; London ; Milan ; Paris ; Singapore ; Tokyo :
Springer, 2000
   (Lecture notes in computer science ; 1812 : Lecture notes in
   artificial intelligence)
   ISBN 3-540-41162-3

CR Subject Classification (1998): I.2.9, I.2.11, I.2.6, I.6

ISBN 3-540-41162-3 Springer-Verlag Berlin Heidelberg New York

Springer-Verlag Berlin Heidelberg New York
a member of BertelsmannSpringer Science+Business Media GmbH
© Springer-Verlag Berlin Heidelberg 2000

Typesetting: Camera-ready by author, data conversion by Steingräber Satztechnik GmbH, Heidelberg
Printed on acid-free paper   SPIN: 10720270   06/3142   5 4 3 2 1 0

Preface

Robot learning is an exciting and interdisciplinary field. This state is reflected in the range and form of the papers presented here. Techniques that have become well established in robot learning are present: evolutionary methods, neural network approaches, reinforcement learning; as are techniques from control theory, logic programming, and Bayesian statistics. It is notalbe that in many of the papers presented in this volume several of these techniques are employed in conjunction. In papers by Nehmzow, Grossmann and Quoy neural networks are utilised to provide landmark-based representations of the environment, but different techniques are used in each paper to make inferences based on these representations.

Biology continues to provide inspiration for the robot learning researcher. In their paper Peter Eggenberger et al. borrow ideas about the role of neuromodulators in switching neural circuits, These are combined with standard techniques from artificial neural networks and evolutionary computing to provide a powerful new algorithm for evolving robot controllers. In the final paper in this volume Bianco and Cassinis combine observations about the navigation behaviour of insects with techniques from control theory to produce their visual landmark learning system. Hopefully this convergence of engineering and biological approaches will continue. A rigorous understanding of the ways techniques from these very different disciplines can be fused is an important challenge if progress is to continue. Al these papers are also testament to the utility of using robots to study intelligence and adaptive behaviour. Working with robots forces us to confront difficult computational problems that may otherwise be all too temptingly swept under the carpet.

In this proceedings we present seven of the talks presented at the 8th European Workshop on Learning Robots in an update and expanded form. These are supplemented by two invited papers, by Ulrich Nehmzow on Map Building for Self-Organisation and by Axel Grossmann and Riccardo Poli on Learning a Navigation Task in Changing Environments by Multi-task Reinforcement Learning. The workshop took place in the friendly surroundings of the Swiss Federal Institute of Technology, (EPFL), Lausanne, Switzerland, in conjunction with the European Conference on Artificial Life '99. The workshop organisers would like to express their deep gratitude to Dario Floreano and all at EPFL who contributed to the success of the meeting. Finally, and certainly not least, we must applaud the programme reviewers who again this year, producing reviews of the highest standard.

June 2000   Jeremy Wyatt
                                                                              John Demiris

Executive Committee

Workshop Chair: Jeremy Wyatt (Birmingham University, UK)
Co-chair: John Demiris (Edinburgh University, UK)

Programme Committee

Minoru Asada (Japan)
Luc Berthouze (Japan)
Andreas Birk (Belgium)
Daniel Bullock (USA)
Kerstin Dautenhahn (UK)
Edwin De Jong (Belgium)
Ruediger Dillmann (Germany)
Marco Dorigo (Belgium)
Dario Floreano (Switzerland)
Luca Maria Gabardella (Switzerland)
Philippe Gaussier (France)
John Hallam (UK)
Inman Harvey (UK)
Gillian Hayes (UK)
Yasuo Kuniyoshi (Japan)

Ramon Lopez de Mantaras (Spain)
Henrik Lund (Denmark)
Sridhar Mahadevan (USA)
Masoud Mohammadian (Australia)
Ulrich Nehmzow (UK)
Stefano Nolfi (Italy)
Ashwin Ram (USA)
Stefan Schaal (USA/Japan)
Patrick van der Smagt (Germany)
Jun Tani (Japan)
Adrian Thompson (UK)
Carme Torras (Spain)
Hendrik Van Brussel (Belgium)

Sponsoring Institutions

MLNet, European Machine Learning Network

Table of Contents

Map Building through Self-Organisation for Robot Navigation

Ulrich Nehmzow

Department of Computer Science
The University of Manchester
Manchester M13 9PL
United Kingdom
ulrich@cs.man.ac.uk

Abstract. The ability to navigate is arguably the most fundamental competence of any mobile agent, besides the ability to avoid basic environmental hazards (e.g. obstacle avoidance).

The simplest method to achieve navigation in mobile robot is to use path integration. However, because this method suffers from drift errors, it is not robust enough for navigation over middle scale and large scale distances.

This paper gives an overview of research in mobile robot navigation at Manchester University, using mechanisms of self-organisation (artificial neural networks) to identify perceptual landmarks in the robot's environment, and to use such landmarks for route learning and self-localisation, as well as the quantitative assessment of the performance of such systems.

1 Introduction

This overview article describes the research in mobile robot navigation conducted at the University of Manchester. It addresses the question of mechanisms for robot map building, mobile robot self-localisation, and landmark selection.

Experimental results are presented, and references to literature giving more details are provided.

2 Motivation

2.1 The Need for Navigation

The ability to move cannot be exploited to its full potential, unless the agent — in this case a fully autonomous mobile robot (see figure 1) — is able to move to specific locations in its environment, that is: to navigate.

Without the ability to navigate, the agent has to resort to strategies of random motion or tropisms, which are often inefficient.

J. Wyatt and J. Demiris (Eds.): EWLR 1999, LNAI 1812, pp. 1–22, 2000.

Fig. 1. *FortyTwo*, the Nomad 200 robot used in the experiments. The robot is equipped with 16 sonar and 16 infrared sensors, flux gate compass and an odometry system. The camera visible in the photograph was not used in the experiments presented here.

2.2   Fundamental Components of Navigation

The ability to navigate entails the following four competences:

1. Map building,
2. self-localisation,
3. map interpretation, and
4. path planning.

For the context of this paper, "map" stands for a one-to-one mapping between physical space (i.e. the world) and map space (i.e. the internal representation of that world). Such a map is not necessarily a metric representation — it can be topological as well. "Map building" refers to the construction of such a mapping, be it the construction of a metric "road map", or some other representation such as an artificial neural network. "Self localisation" refers to the navigator's ability to make the connection between the world and its representation, i.e. to establish his position in the map. Finally, "map interpretation" refers to the use of that mapping for navigational purposes such as homing, route following, or in general path planning of arbitrary paths.

If any of these four competences is missing, full navigation (i.e. movement to an arbitrary goal) becomes impossible, and the navigator has to resort to either random movement (search behaviour), or tropism (moving towards an attractor).

2.3   Methods of Navigation

Broadly speaking, there are two major options to achieve robot navigation. One is to use proprioceptive sensors such as wheel encoders, and to perform navigation through path integration (also often referred to as dead reckoning). The other option is to use exteroceptive sensors, and to navigate using landmarks.

Dead reckoning. Dead reckoning allows the navigator to determine its current position within the map, provided the following information is available:

- Map,
- known starting location,
- current heading, and
- either current speed or distance travelled.

Starting from a known position (x_0, y_0), the navigator measures the distance travelled and the current heading. Using trigonometric relationships, the respective movement in x and y direction can be determined, and an updated estimate of the navigator's current position computed.

The major problem with navigation by path integration is that distance travelled and, even more importantly, heading cannot be measured perfectly, but are subject to an unknown error. This leads to the accumulation of a drift error, which can only be corrected by referring to external information, for example landmarks.

Landmark based navigation. An alternative method is to navigate with reference to perceptual landmarks that are detectable by exteroceptive sensors. Provided landmark recognition is unambiguous, this method does not suffer from the accumulation of incorrigible drift errors.

However, if the recognition of landmarks is not unambiguous ("perceptual aliasing"), navigation can become unreliable, and additional methods have to be sought to establish the navigator's position unambiguously. This is discussed later in section 4.2.

2.4   Why Self-Organisation Is a Good Idea

I argue that there are two reasons why mechanisms of self-organisation (i.e. learning) should be used in preference to pre-installed, fixed mechanisms in autonomous mobile robot navigation. The first one is a methodological consideration, the second reason has to do with the nature of the robot's perception of its environment.

1. Navigation is a core competence of all moving animals, and therefore plays a very important role in our lives. We perform navigation tasks so routinely that we are usually not even aware of the fact that we *are* navigating. Only when things get difficult — for example when we are lost or moving through unknown terrain — we speak of "navigating". But we are, in fact, navigating all the time when moving, making decisions as to which way to turn (path planning), updating our representation of space (map building), etc.

Because of this dominant, yet concealed role of navigation, it is virtually impossible to avoid making assumptions regarding *robot* navigation that are based on *our* experience. And yet, there are such enormous perceptual and behavioural differences between humans and robots that it is likely that these assumptions are ill founded. A typical example of this phenomenon is the choice of landmarks in an office environment. Almost invariably, humans rank doors very highly as landmarks suitable for navigation. It turns out, however, that for robots using sonar sensors — one of the most widely used sensor modalities in mobile robotics — doors themselves are not that easy to detect. Door *frames*, on the other hand, show up very clearly on a sonar scan.

The first argument, therefore, for using self-organisation in mobile robot navigation is that it can reduce the amount of predefinition put in by a human operator.

2. Robot sensors are subject to noise. No two measurements of the same object are identical. Furthermore, the sensory perception of a robot can be plain wrong and therefore misleading (e.g. through specular reflection of a sonar signal), contradictory (e.g. in information regarding the same object, coming from different sensors), or just useless (e.g. information from sensors that do not supply any meaningful information with respect to the task being performed by the robot).

To make sense of noisy, inconsistent, contradictory or useless data is a hard problem, in particular if the data is to be used in a clearly defined, crisp model.

Learning mechanisms of self-organisation make use of the data that is actually available, without prior assumptions. This takes care of useless data. Self-organising mechanisms such as artificial neural networks can cluster information topologically, which addresses the problem of noise. And finally, if there is information in the sensor signal that allows to detect inconsistency or contradiction (e.g. if one sensor produces a freak signal, but all neighbouring sensors provide the correct reading), mechanisms of self-organisation have proven to be a good method of eliciting that information.

The second argument in favour of self-organisation, therefore, is that such mechanisms are well suited to process the kind of data obtained from robot sensors.

3   Clustering Mechanisms

3.1   Self-Organising Feature Map

The self-organising feature map (SOFM), or Kohonen network, is a mechanism that performs an unsupervised mapping of a high dimensional input space onto a (typically) two-dimensional output space ([Kohonen, 1988]).

The SOFM normally consists of a two-dimensional grid of units, as shown in figure 2.

All units receive the same input vector $\imath$. Initially, the weight vectors $\boldsymbol{w}_j$ are initialised randomly and normalised to unit length.

The output o_j of each unit j of the net is determined by equation 1.

$$o_j = \boldsymbol{w}_j \cdot \imath. \tag{1}$$

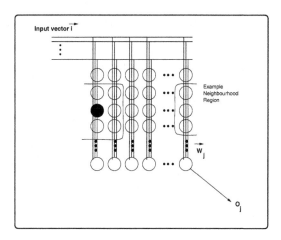

Fig. 2. Self-organising feature map.

Because of the random initialisation of the weight vectors, the outputs of all units will differ from one another, and one unit will respond most strongly to a particular input vector. This "winning unit" and its surrounding units will then be trained so that they respond even more strongly to that particular input vector, by applying the update rule of equation 2. After having been updated, weight vectors w_j are normalised again.

$$w_j(t + 1) = w_j(t) + \eta(\imath - w_j(t)), \tag{2}$$

where η is the learning rate (usually a value around $\eta = 0.3$). The neighbourhood around the winning unit, within which units get updated, is usually chosen to be large in the early stages of the training process, and to become smaller as training progresses. Figure 2 shows an example neighbourhood of one unit around the winning unit (drawn in black). The figure also shows that the network is usually chosen to be torus-shaped, to avoid border effects at the edges of the network.

As training progresses, certain areas of the SOFM become more and more responsive to certain input stimuli, thus clustering the input space onto a two-dimensional output space. This clustering happens in a topological manner, mapping similar inputs onto neighbouring regions of the net.

3.2 Reduced Coulomb Energy Networks

The Reduced Coulomb Energy (RCE) Classifier ([Reilly et al., 1982]) is another method of classification. Examples of its application to mobile robotics are [Kurz, 1994,Kurz, 1996] and [Nehmzow and Owen, 2000].

In the RCE net, each class is represented by a *representation vector* (R-vector). Training the RCE-Classifier involves determining the R-vectors.

When a pattern is presented to the classifier, the input is compared to each of the already existing R-vectors, using some form of similarity measure (e.g. dot product) in order to determine the R-vector of highest similarity with the currently perceived pattern. If the similarity between the input pattern and the "winning" R-vector is within a pre-determined threshold then the input pattern belongs to the class of this winning R-vector. If the similarity is outside the threshold then the input pattern becomes a new R-vector. Thus the boundaries of classes are determined by the nearest neighbour law. Figure 3 shows an example for a two-dimensional input vector.

Because of this learning procedure, the RCE network grows in size as more and more input stimuli are presented.

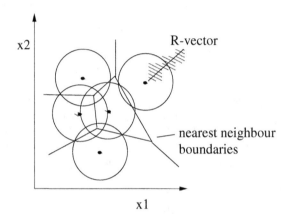

Fig. 3. RCE-Classifier, two-dimensional example. Each "dot" in the diagram represents an R-vector. The circles surrounding each R-vector denote the "threshold area" within which an input pattern must fall in order to belong to the corresponding R-vector's class. In the case of patterns falling within more than one threshold area, the nearest neighbour law applies.

3.3   Adaptive Resonance Theory

A third example of a self-organising artificial neural network is the Adaptive Resonance Theory (ART) network. The basic ART architecture consists of two fully connected layers of units, a feature layer (F1) which receives the sensory input, and a category layer (F2) where the units correspond to perceptual clusters or prototypes ([Grossberg, 1988], see figure 4). There are two sets of weights between the layers, corresponding to feedforward and feedback connections. A "winner-takes-all" criterion is used during the feedforward phase, and a similarity criterion is used to accept or reject the resulting categorisation in the feedback phase.

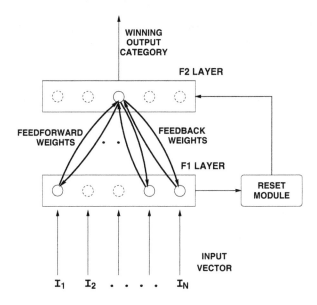

Fig. 4. Basic ART Architecture. The units in the feature layer (F1) are completely connected to each of the units in the category layer (F2), through two separate sets of weights. For clarity, only a few of the feedforward and feedback connections are shown.

When an input pattern is presented to the network, it is compared to each of the existing prototypes (through the feedforward weights) to determine a winning node. If the similarity between the input pattern and the winning node (through the feedback weights) exceeds a vigilance threshold, then adaptive learning is activated and the stored pattern is modified to be more similar to the input pattern (the learning method depends on the particular variant of ART network used). Otherwise a "reset" occurs, where the current winner is disabled, and the network searches for another node to match the input pattern. If none of the stored prototypes is similar enough to the given input, then a new prototype will be created corresponding to the input pattern. The ART network, therefore, grows like the RCE network as new data is presented to the net.

4   Applications to Mobile Robotics

4.1   Route Learning

Mechanism. Self-organising feature maps can be used to achieve route learning in a mobile robot ([Nehmzow and Owen, 2000]). In experiments conducted in Manchester, route following was accomplished by learning perception-reaction pairs that resulted in the robot staying on the trained route. Provided perceptual aliasing (the confusion of locations because of their perceptual similarity) is either not present at all or low, the association of perception with action will successfully train a robot to follow a user-defined route. At Manchester, routes of over 100 m length are now routinely trained and learned. One such route is shown in figure 5.

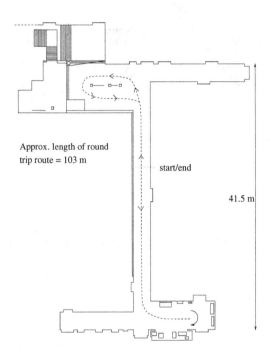

Approx. length of round
trip route = 103 m

start/end

41.5 m

Fig. 5. A route of about 100 m learned by *FortyTwo*.

To achieve noise resilience on the one hand, and sensor-motor association on the other, a self-organising feature map was used as an associator (see figures 2 and 6).

During a user-supervised training phase, sensor signals (11 sonar readings and 11 infrared readings) as well as user steering commands (a 2-bit encoding of motor commands) form the input to a SOFM. This means that due to the learning mechanism of the SOFM, perception-action pairs are stored in the weight vector of each unit of the network.

During *autonomous* traversal of the trained route, the motor action component of the SOFM is set to zero, and the winning unit of the network is determined using sensory perception alone. The winning unit's weight vector is then used to determine the motor action required to stay on the trained route. This mechanism is shown in figure 6.

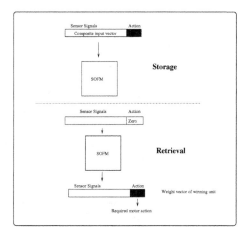

Fig. 6. SOFM as associator of perception-action pairs for route learning.

Results. *FortyTwo* was successfully trained to learn routes within laboratory environments and in the corridors of the Department of Computer Science ([Nehmzow and Owen, 2000,Nehmzow, 1999]). Small deviations from the trained route due to perceptual aliasing do occur, but our observation was that the robot always encountered a unique sensory perception soon enough to regain the correct route. In our experiments, 11% of the network's cells fired in more than on location, i.e. were subject to perceptual aliasing. But, as said above, this did not prevent the robot from learning routes successfully.

Extension to point to point navigation. That such a route learning system can be extended to allow navigation between arbitrary locations within a mapped area was subsequently demonstrated in experiments discussed in detail in [Nehmzow and Owen, 2000] and [Owen, 2000].

In those experiments, a topological map based on a RCE network was extended to include direction and distance information between landmarks — the so-called vector map. In conjunction with additional enabling competences (wall following, centring at a location, and returning to a previous location on the route), *FortyTwo* was able to determine *arbitrary* routes within the Computer Science Department at Manchester, and to follow them successfully.

4.2   Self-Localisation

Any map is meaningless if the navigator's position within the map is unknown. The first step towards map-based navigation, therefore, is self-localisation.

Experiments carried out in Manchester ([Duckett and Nehmzow, 1996] and [Duckett and Nehmzow, 1998]) address the problem of localisation in autonomous mobile robots in environments that show a high degree of perceptual aliasing. In particular, the work was concerned with the more general problem of re-localisation, where the robot starts from an unknown location within the mapped area, and aims to establish its position using perceptual cues as well as information about its egomotion.

Underlying mechanism. The fundamental mechanism used for self-localisation is best explained using an example from [Duckett and Nehmzow, 1996]. This work was later extended, and is described in [Duckett and Nehmzow, 1998] and [Duckett, 2000].

The architecture of the localisation system is shown in figure 7.

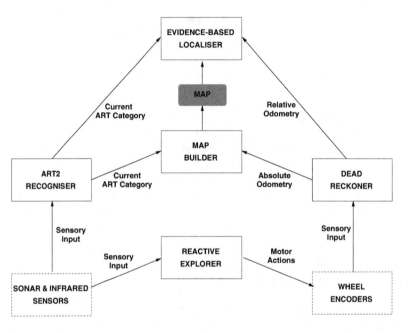

Fig. 7. Localisation System Architecture. The solid-edged boxes denote behavioural modules, and arrows the dependencies between different parts of the system. The shaded box denotes the representation used for the map, and the dashed boxes show hardware components.

During an exploration phase, the robot builds a map of its environment. This is done by using an artificial neural network such as ART. Other classification mechanisms that categorise sensory perception can also be used (see, for instance, [Duckett and Nehmzow, 1998]). During the entire operation of the robot, i.e. during the map building phase and during the localisation phase, the orientation of the robot's turret is kept constant, so that sensory perceptions are only dependent upon the robot's location, not its heading.

The robot is then moved to a randomly chosen position within that environment, where it will attempt to localise. By actively exploring, and accumulating evidence through the use of relative odometry between local landmarks, the robot is able to determine its location with respect to perceptual landmarks very quickly.

Localisation algorithm. Possible locations within the map are stored as a list of hypotheses in working memory, each with an associated confidence level. Whenever the robot perceives a new location, either by recognising a change in (clustered) sensory perception or a change in odometry bigger than $2D$ (where D is a preset distance threshold) another set of possible locations is created. A matching procedure is then used to combine both sources of evidence to produce an updated list of hypotheses. The algorithm is given as follows, and explained in more detail below.

0. *Initialisation.* Formulate set of hypotheses, $H = \{h_0, ..., h_N\}$, consisting of location points which match the current ART category. Initialise confidence levels: $\forall h_i \in H$, assign $conf(h_i) = 1$.
1. Wait until ART category changes or robot has moved distance $2D$, where D is the distance threshold used during map building.
2. For each hypothesis h_i add change in odometry $(\triangle x, \triangle y)$ to the coordinates of h_i, (x_{h_i}, y_{h_i}).
3. Generate second set of candidate hypotheses, $H' = \{h'_0, ..., h'_N\}$.
4. For each h_i, find the nearest h'_j, recording distance apart, d_{h_i}.
5. Accumulate evidence, given distance threshold T, minimum confidence level MIN, gain factor $GAIN > 1$ and decay factor $0 < DECAY < 1$:
   $\forall h_i \in H$
       if $d_{h_i} < T$ then
           replace (x_{h_i}, y_{h_i}) with $(x_{h'_j}, y_{h'_j})$ from the matching h'_j
           let $conf(h_i) = conf(h_i) \times GAIN$
       else
           let $conf(h_i) = conf(h_i) \times DECAY$
           if $conf(h_i) < MIN$ then delete h_i.
6. Delete any duplicates in H, preserving the one with the highest confidence level.
7. Add all remaining h'_j from H' which are not already contained in H to H, assigning $conf(h'_j) = 1$.
8. Repeat from step 1.

In step 2, the existing set of location estimates are updated by adding on the change in position recorded by relative odometry. A new set of candidate hypotheses is then assembled by selecting all possible location points from the map which match the current ART category. A search procedure follows, where each of the existing hypotheses is matched to its nearest neighbour in the new candidate set. Step 5 then uses a threshold criteria to determine whether each of the matched hypotheses should be increased or decreased in confidence. This means that if the matched position estimates are sufficiently close to each other, then this is judged to be a piece of evidence in favour of this particular hypothesis. The confidence level is therefore raised by a gain term, and the position estimate from the old hypothesis is replaced by the new value. Thus, good position estimates are continually corrected on the basis of perception.

Conversely, if the match distance exceeds the threshold, then the associated confidence level is scaled down by a decay term, and the position estimate is left uncorrected, i.e, it is not replaced by its nearest neighbour in the candidate set. Hypotheses which fall below a certain confidence level are rejected and pruned from the list of possible locations, eliminating the bad hypotheses quickly and minimising the search space. Duplicates created by the matching process are deleted here too.

Any unmatched location points remaining in the candidate set are also added to the current list of hypotheses, and are assigned an initial confidence value. This will be the case for all of the candidate hypotheses on the first iteration, as the hypothesis set will initially be empty. (Initialisation has been included as step 0 here for clarity, although this is actually unnecessary.) All possible locations with the current perceptual signature are always considered, so that the algorithm can cope with unexpected, arbitrary changes of position.

Eventually, one of the hypotheses emerges as a clear winner. For example, in experiments conducted on *FortyTwo* in our laboratory, this took an average of 7 iterations through the algorithm, in a mean time of 27 seconds with the robot travelling at 0.10 ms^{-1}. If the robot becomes lost again, confidence in this particular hypothesis gradually decays as a new winner emerges.

Figure 8 shows a map of our laboratory environment. The underlying reference frame of this map is a Cartesian space. The clusters indicate the size of areas within which the robot's ART map perceives the same landmark, and numbers within the clusters indicate the ART category being perceived at that location. It is obvious from that map that perceptual aliasing occurs, because there are several locations that all elicit the same response from the network.

Results. The system was tested on *FortyTwo* in the robotics laboratory at Manchester University. A wall following behaviour was used for exploration, and an ART2 network configured to receive input from the robot's 16 infra-red sensors. Map building was carried out during a single circuit of an enclosure made from walls and boxes. This was purposely designed to contain areas of perceptual aliasing, as shown in the map (figure 8). The average forward speed of the robot travelling around this enclosure was 0.10 ms^{-1}.

The algorithm was run for 20 iterations per trial for 30 trials. The results given in figure 9 show a steady downward trend in the mean localisation error over time. The "actual" position of the robot had to be taken from global odometry, because no external means of recording *FortyTwo's* real position is currently available. The results shown are therefore only as accurate as the drift in odometry will allow.

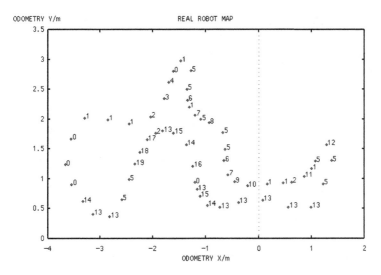

Fig. 8. Map of a laboratory environment in Manchester. Numbers indicate the perceptual category identified by the ART network, locations of these perceptual clusters are given in Cartesian coordinates.

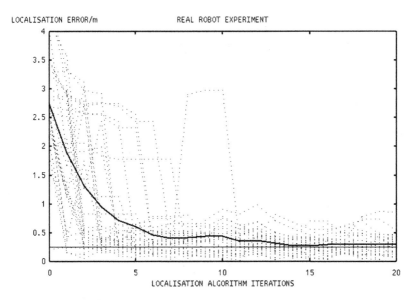

Fig. 9. *Localisation Error on Real Robot.* The thick line shows the mean error taken over 30 trials. The dotted lines show the individual trials, and the horizontal line at $y = 0.25$ denotes the distance threshold D used during map building.

In addition, the time and number of iterations through the algorithm taken for the localisation error to fall below the distance threshold D was measured in each trial. The mean number of steps was 7.0, with a standard deviation of 5.6. The mean time taken was 27.3 seconds, with a standard deviation of 21.9 seconds. (The average curve shown in figure 9 takes longer than this to reach the D level because this calculation takes into account the worst case localisation errors.)

Inspection of individual localisation attempts showed that the algorithm was able to localise successfully once the robot had travelled through a unique sequence of perceptions along the perimeter of the enclosure. For example, following the route taken by the robot clockwise in figure 8, the sequence "0, 1, 5" of ART2 categories uniquely identifies the point at (-1.28, 2.81). However, the sequence "0, 1" would be ambiguous, because there are two possible places in the route where this occurs.

Quantitative localisation performance assessment. While it was possible to make *qualitative* judgements about the behaviour of the localisation system on the real robot, the use of global odometry — which is subject to incorrigible drift errors — for calculating the localisation error makes it very difficult to make a really accurate assessment on the performance of the localisation algorithm. We have therefore introduced quantitative localisation performance measures which are independent from robot odometry. These are discussed in the next section.

5   Assessing Localisation Performance Using Contingency Table Analysis

5.1   Contingency Table Analysis

In order to assess the performance of the localisation system, we collected location-response pairs of the localisation system in a contingency table like the one shown in figure 10.

Such a contingency table states how many times a particular localisation system response was obtained at the various physical locations the robot occupied during its operation[1]. Contingency table analysis can then be used to determine whether there is a statistically significant relationship between localiser response and position of the robot ([Press et al., 1992,Nehmzow, 1999]).

Particularly useful in this respect is the uncertainty coefficient U, defined by equation 3.

$$U(L \mid R) \equiv \frac{H(L) - H(L \mid R)}{H(L)}, \tag{3}$$

with the respective entropies defined as $H(L \mid R) = H(L, R) - H(R)$, $H(L, R) = -\sum_{r,l} p_{rl} \ln p_{rl}$ and $H(L) = -\sum_{l} p_{\cdot l} \ln p_{\cdot l}$.

---

[1] The physical environment was compartmentalised into distinct zones for the purpose of indicating the robot's current position in the world.

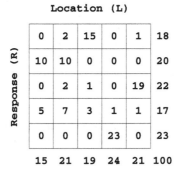

Location (L)

	0	2	15	0	1	18
Response (R)	10	10	0	0	0	20
	0	2	1	0	19	22
	5	7	3	1	1	17
	0	0	0	23	0	23
	15	21	19	24	21	100

Fig. 10. Example Contingency Table. The rows correspond to the response produced by the localisation system, and the columns to the "true" location of the robot as measured by an observer. This table represents 100 data points, and also shows the totals for each row and column.

The probabilities are defined by equations 4 and 5.

$$p_{.l} = \frac{N_{.l}}{N},\tag{4}$$

$$p_{rl} = \frac{N_{rl}}{N},\tag{5}$$

and the column totals $N_{.l}$ and the table total N are given by equations 6 and 7 respectively.

$$N_{.l} = \sum_{r} N_{rl},\tag{6}$$

$$N = \sum_{r,l} N_{rl}.\tag{7}$$

The uncertainty coefficient U takes a value between 0 ("no correlation between location and localiser response") and 1 ("perfect localisation").

5.2 Application to Robot Self-Localisation

We used the uncertainty coefficient U, as introduced in the previous section, to compare the localisation performance of three different localisation mechanisms, in six different real world environments.

The three mechanisms used were

1. Dead reckoning, using the robot's on-board odometry system,
2. landmark-based localisation, using sonar and infrared sensor information clustered by a nearest neighbour classifier, and

3. evidence based localisation, using sonar and infrared sensor information, as well as local odometry information and the change of perception as known movements are executed (for details see [Duckett and Nehmzow, 1998]).

The six environments in which these three localisers were compared are shown in table 1.

Table 1. Characterisation of the six different environments in which entropy-based performance measures were applied to the robot's evidence-based localisation (EBL) system.

Description	Route Location in m	No. of Bins	Trials	Places in EBL Map
A Drinks-machine area	60	24	298	88
B T-shaped hallway	54	14	263	71
C L-shaped corridor	146	40	474	185
D Small empty room	23	8	232	33
E Single corridor	51	14	248	61
F E plus moving people	51	14	249	61

Table 2 shows the localisation results obtained using localisation by dead reckoning, localisation using a landmark classifier based on a self-organising feature map, and localisation using the evidence-based localiser discussed in section 4.2.

Table 2. Summary of Results for evidence-based localisation, localisation by perceptual landmarks, and localisation by uncorrected dead reckoning (cf. figure 11). $\overline{U(L \mid R)}$ is the mean uncertainty coefficient.

	EBL Dist./m for $U(L \mid R) = 0.9$	Landmark Classifier $\overline{U(L \mid R)}$	Dead Reckoning $\overline{U(L \mid R)}$
A	0.0	0.925	0.663
B	3.8	0.814	0.724
C	13.6	0.739	0.785
D	0.5	0.878	0.786
E	3.0	0.776	0.864
F	4.7	0.690	0.686

Table 2 and figure 11 show how evidence-based localisation consistently provides the best localisation results, and how in dead reckoning localisation performance decreases as the travel distance increases. Localisation purely based on landmarks shows consistent performance versus distance travelled, as it uses only perceptual information and does not take temporal information into account.

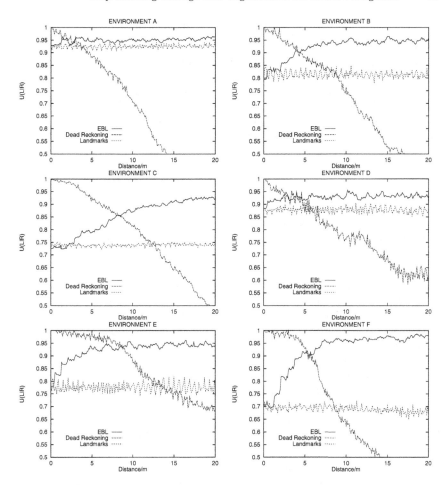

Fig. 11. Localisation results in the six different environments. In all environments local-isation by dead reckoning deteriorates with increasing travel distance, while localisation using evidence (EBL) improves. Localisation based purely on landmark recognition is not affected by distance travelled ($U(L \mid R) = 1$ for perfect localisation).

6   Open Questions and Ongoing Research

Having developed mobile robot navigation mechanisms that allow mobile robots to learn routes, to localise, and to plan paths within middle scale environments of several hundred m^2 size, we are currently looking at mechanisms that make these navigation systems more efficient.

6.1   Learning Parameters and Network Structure

For robot applications in potentially unbounded environments, for example for applications of continuous operation ("lifelong learning"), networks with a fixed capacity (such as the self-organising feature map) will eventually saturate. At that point, it is impossible to store more landmarks without erasing some information from the net.

We have therefore used growing networks such as ART and RCE, and are currently investigating modifications to the self-organising feature map that would also address the problem of saturation. This work, however, is still in its early stages.

6.2   Landmark Selection

One possible method increasing the efficiency of a map-based navigation system is not to store *every* perceptual landmark encountered, but to make a selection. One possible method of making such a selection is novelty detection.

Novelty detection. The primary purpose of our research in novelty detection for mobile robotics is to detect deviations from a norm by means of acquiring a "standard model" of the robot's environment, and to flag deviations from that model that exceed a predefined threshold. Possible applications include monitoring tasks, whose objective it is to detect faults and critical deviations from the norm in the robot's environment.

Such a novelty detection system can be used for landmark selection, too, by eliminating common perceptual landmarks, i.e. those landmarks which say little about the robot's current position.

[Marsland et al., 2000] describe the application of a novelty detector to identifying novel stimuli impinging on a robot's sensor. To achieve this, sensory perceptions (one of four perceptually unique light stimuli) were first clustered using either a standard self-organising feature map (see figure 2), or a modified self-organising feature map with leaky integrators, i.e. a network whose neuron activity decays over time.

The output of the winning unit of the network was then fed into a habituating output neuron, whose synaptic efficacy $y(t)$ decreased according to equation 8.

$$\tau \frac{dy(t)}{dt} = \alpha \left[y_0 - y(t) \right] - S(t), \tag{8}$$

The effect of this equation is shown in figure 12.

The return to high synaptic efficacy at time 150 in figure 12 is due to a dishabituation behaviour, which can be activated or de-activated according to the application's needs.

Initial experiments show that habituating, self-organising maps such as these can be used for landmark selection as well. Only landmarks that present "rare" stimuli pass the filter, reducing the number of stored landmarks per map area considerably. Again, the application of landmark selection through novelty filtering is subject to ongoing research.

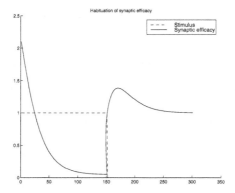

Fig. 12. An example of how the synaptic efficacy drops when habituation occurs (equation 8). Both stimulus and synaptic efficacy are shown: for $0 < t < 150$ the stimulus is constantly "1", then drops to "0".

7   Summary and Conclusion

7.1   Self Organisation for Map Building, Route Learning and Self-Localisation

The ability to navigate is one of the most important competences for any moving agent. Without purposive motion, the advantages of mobility cannot be fully exploited.

One of the core components of a navigational skill is the ability to construct internal representations ("maps") of the world. A map can then be used for navigation if the agent has the additional competences of self localisation within the map, map interpretation, and path planning.

This paper presents two examples of how map building in a mobile robot can be achieved through self organisation. Using a self organising mechanism has the double advantage of minimising the use of user-defined assumptions, and of taking the *actual* sensory perceptions into account when map building, rather than abstracted model perceptions. This increases the navigator's resilience to noise.

Route learning and point to point navigation. In a first robot navigation example, a self-organising feature map — an artificial neural network that clusters the robot's sensory perceptions topologically — was used in a route learning and free navigation system. During a supervised training phase, the network was trained, using sensory perceptions as well as steering commands as inputs. In this manner, perceptions were associated with motor actions.

During the subsequent autonomous route following phase, sensory perceptions alone were used to determine which of the network's artificial neurons responded most closely to that particular sensory stimulus. The motor action component of the weight vector of that neuron then identified the required motor action.

Self localisation and localisation performance assessment. A second example showed how an Adaptive Resonance Theory self-organising network can be used to establish the robot's position within the map, having started from an unknown location. This is the "lost robot problem".

Integrating both perceptual and temporal information, the robot constructs a list of map locations that have the same perceptual signature as the robot's current location in the world. Taking into account the robot's egomotion as determined through dead reckoning, the list of hypotheses is continuously shortened, until only one candidate location remains. At this point, self-localisation is achieved.

In our experiments in Manchester, our mobile robot *Forty Two* was *always* able to self-localise. The travel distance this would take was dependent on the environment the robot operated in, but was typically 2 to 5 m of travel, and hardly ever exceeded 10 m.

To assess the performance of our localisation system and to compare it with other localisation methods, we used contingency table analysis, which allowed us to grade the localisation performance *quantitatively*, without using drift-affected odometry information at all.

7.2   Problems of Map Building through Self-Organisation

We identified two major problems with the approach to map building and navigation through topological clustering of sensory information, and proposed solutions to these problems, too.

Perceptual aliasing. Constructing maps from topologically clustered sensory perceptions will result in a certain degree of perceptual aliasing (in our route learning experiments, a little over 10% of all network units fired in more than one location).

Increasing the sensor resolution of the robot might reduce this number, but may not lead to better navigation systems, because very high sensor resolution would reduce the recognition rate of landmarks on subsequent visits. It would also increase the information processing burden as additional, often irrelevant information would have to be processed.

The way we addressed the problem of perceptual aliasing is discussed in section 4.2. Retaining the relatively low sensor resolution of the robot, we differentiated between two locations that looked alike by taking into account the path (i.e. the temporal pattern of perceptions) by which the robot arrived at the current location.

Saturation. Unless the self-organising map building mechanism is to be used in a closed environment of limited size, the problem of network saturation becomes relevant.

If a network of finite, constant size such as the self-organising feature map is used, there will be a point at which further landmark information can only be stored in the net if previously stored information is removed. This is usually undesirable.

To address this problem, we used growing neural networks such as the Restricted Coulomb Energy network (section 3.2) or Adaptive Resonance Theory (section 3.3). Both networks add new neurons when novel information is encountered. One problem in this respect is to find the right balance between generalisation and specificity, i.e. to determine the parameter that controls the adding of new neurons. We are not aware of a principled approach to find the right parameter, and have used trial and error in our experiments.

7.3   Focusing the Stored Information through Landmark Selection

One approach to increasing the information content of the map that we consider promising is that of landmark selection. The idea is that instead of storing *all* sensory perceptions that the robot encounters during the mapping phase, only those perceptual landmarks are stored that differ significantly from commonly encountered landmarks.

Initial experimental results of using a habituating self-organising map to act as a landmark selector, a novelty filter, have proven promising, and we believe that the addition of a landmark selection filter prior the map builder (section 6.2) will improve the performance and efficiency of landmark based navigation systems.

Acknowledgements

The work described in this article is a team effort, and I acknowledge the valuable contribution of my co-workers.

The work on route learning is Carl Owen's PhD research work, the work on self-localisation is part of Tom Duckett's PhD research. Jonathan Shapiro and Stephen Marsland are the co-workers in novelty detection.

The work reported here was made possible through financial support by the Department of Computer Science at the University of Manchester. Carl Owen and Stephen Marsland were supported by EPSRC PhD studentships, and Tom Duckett was supported by a Departmental studentship.

References

Duckett, 2000. Duckett, T. (2000). *Concurrent Map Building and Self-Localisation for Mobile Robot Navigation*. PhD thesis, Department of Computer Science, University of Manchester. to appear.

Duckett and Nehmzow, 1996. Duckett, T. and Nehmzow, U. (1996).   A robust perception-based localisation method for a mobile robot.   Technical Report UMCS-96-11-1, Department of Computer Science, University of Manchester.

Duckett and Nehmzow, 1998. Duckett, T. and Nehmzow, U. (1998).   Mobile robot self-localisation and measurement of performance in middle scale environments. *Robotics and Autonomous Systems*, 24(1–2):57–69.

Grossberg, 1988. Grossberg, S. (1988). *Neural Networks and Natural Intelligence*. MIT Press, Cambridge MA.

Kohonen, 1988. Kohonen, T. (1988). *Self Organization and Associative Memory*. Springer Verlag, Berlin, Heidelberg, New York.

Kurz, 1994. Kurz, A. (1994). *Lernende Steuerung eines autonomen mobilen Roboters*. PhD thesis, Fachbereich 19, Technische Hochschule Darmstadt.

Kurz, 1996. Kurz, A. (1996). Constructing maps for mobile robot navigation based on ultrasonic range data. *IEEE Transactions on Systems, Man, and Cybernetics–Part B: Cybernetics*, 26(2):233–242.

Marsland et al., 2000. Marsland, S., Nehmzow, U., and Shapiro, J. (2000). Novelty detection for robot neotaxis. In *Proc. NC'2000, Second International ICSC Symposium on NEURAL COMPUTATION, Berlin*.

Nehmzow, 1999. Nehmzow, U. (1999). *Mobile Robotics: A Practical Introduction*. Springer Verlag, Berlin, Heidelberg, New York.

Nehmzow and Owen, 2000. Nehmzow, U. and Owen, C. (2000). Robot navigation in the real world: Experiments with Manchester's FortyTwo in unmodified, large environments. *J. Robotics and Autonomous Systems*, 33. to appear.

Owen, 2000. Owen, C. (2000). *Map-Building and Map-Interpretation Mechanisms for a Mobile Robot*. PhD thesis, Department of Computer Science, University of Manchester. to appear.

Press et al., 1992. Press, W., Teukolsky, S., Vetterling, W., and Flannery, B. (1992). *Numerical Recipes in C*. Cambridge University Press, Cambridge.

Reilly et al., 1982. Reilly, D., Cooper, L., and Erlbaum, C. (1982). A neural model for category learning. *Biological Cybernetics*, 45:35–41.

Learning a Navigation Task
in Changing Environments
by Multi-task Reinforcement Learning

Axel Großmann and Riccardo Poli

School of Computer Science, The University of Birmingham,
Birmingham, B15 2TT, UK

Abstract. This work is concerned with practical issues surrounding the
application of reinforcement learning to a mobile robot. The robot's task
is to navigate in a controlled environment and to collect objects using
its gripper. Our aim is to build a control system that enables the robot
to learn incrementally and to adapt to changes in the environment. The
former is known as multi-task learning, the latter is usually referred to
as continual 'lifelong' learning. First, we emphasize the connection between adaptive state-space quantisation and continual learning. Second,
we describe a novel method for multi-task learning in reinforcement environments. This method is based on constructive neural networks and
uses instance-based learning and dynamic programming to compute a
task-dependent agent-internal state space. Third, we describe how the
learning system is integrated with the control architecture of the robot.
Finally, we investigate the capabilities of the learning algorithm with respect to the transfer of information between related reinforcement learning tasks, like navigation tasks in different environments. It is hoped that
this method will lead to a speed-up in reinforcement learning and enable
an autonomous robot to adapt its behaviour as the environment changes.

1 Introduction

1.1 Reinforcement Learning for Robots

It is desirable that our robots learn their behaviours rather than have them hand-
coded. This is the case because programming all the details of the behaviours
by hand is tedious, we may not know the environment ahead of time, and we
want the robots to adapt their behaviours as the environment changes.

Reinforcement learning (RL) [10] has been used by a number of researchers
as a computational tool for building robots that improve themselves with experience [11, 16, 17, 19, 27]. Strictly speaking, reinforcement learning is a problem
formulation. It defines the interaction between a learning agent and its environment in terms of states, actions, and rewards. A reinforcement-learning agent
improves its performance on sequential tasks using reward and punishment received from the environment. There is no teacher that tells the agent the correct
response to a situation when the agent performs poorly. An agent's only feedback, the only indication of its performance, is a scalar reward value. The task of

J. Wyatt and J. Demiris (Eds.): EWLR 1999, LNAI 1812, pp. 23–43, 2000.

the learning algorithm is to find a policy, mapping environment states to actions, that maximises the reward over time.

Reinforcement learning methods try to find an optimal policy by computing a cumulative performance measure from immediate rewards using statistical methods and dynamic programming. There are a number of reinforcement learning techniques that work effectively on a variety of small problems [1, 24, 32]. But only few of them scale up well to larger problems. As generally known, it is difficult to solve arbitrary problems in the general case. Kaelbling *et al.* [10] suggested therefore to give up *tabula rasa* learning techniques and to guide the learning process by shaping, local reinforcement signals, imitation, problem decomposition, and reflexes. These techniques incorporate a search bias into the learning process which may lead to speed-ups.

Apart from the scalability problem, which is independent of the application domain, additional difficulties arise if reinforcement learning methods are applied in robotics. Such problems are the existence of hidden state, the slow convergence in stochastic environments with large state spaces, and the need for generalisation.

In general, a robot will have limitations in observing its environment. Some uncertainty will remain in the agent's belief about the current state of the world. The usual approach is to use memory of previous actions and observations to aid the disambiguation of the states in the world. In this work, we assume that a method for state disambiguation is available. Although there may be a great deal of uncertainty about the effects of an agent's actions, there is never any uncertainty about the agent's current state.

1.2 Speeding Up the Learning Process by Incremental Learning

Ideally, we would like to build mobile robots that learn incrementally and that adapt to changes in their environment during their entire lifetime. Hopefully, this will also speed up the learning process. In most RL work to date, the algorithms do not use previously learnt knowledge to speed up the learning of a new task. Instead, the learning agent either ignores the existing policy, or worse, the current policy harms the agent while learning the next one. Therefore, we would like to address the issue of learning multiple tasks in reinforcement environments.

The main challenge in building agents capable of incremental learning is to find effective mechanisms for the transfer of information between learning tasks. Recently, this need has been recognised in the machine learning community by initiating a new research direction, termed 'learning to learn' [31]. Transfer in supervised learning involves reusing the features developed for one classification or prediction task as a bias for learning related tasks [3, 4, 30]. Transfer in reinforcement learning involves reusing the information gained while learning to achieve one goal for learning to achieve other goals more easily [25, 30]. At the time of writing, only very few of these approaches have been applied to real robots [30].

Ring [25] defined continual learning as incremental, hierarchical development. He suggested that an agent should make use of previously learnt information when it has to learn a new behaviour. Moreover, the behaviours should be learnt in a bottom-up process whereby the old behaviours are used as constituents of newly created behaviours.

In this work, we embrace these ideas and focus on two important problems of continual learning: the adaptability to changes in the environment and multi-task learning. First, a continual-learning agent should be able to cope with an environment that changes unpredictably with time. It should adjust its behaviour to both gradual and sudden change [20]. Second, the agent should learn incrementally. That is, in learning multiple tasks, it should improve its performance at each task with experience and with the number of tasks [30]. For an algorithm to fit this definition, some kind of transfer must occur between multiple tasks. This means that the representations used by the learning method must support incremental learning. Tree-structured representations [18] and constructive neural networks do so and therefore can solve these two problems. Here, we use connectionist function approximators. Since incremental learning in neural networks using distributed representations is still an open issue, we decided to use neural networks that are based on local representations only.

The aim of this work is to make a contribution towards the development of continual learning methods *applicable to mobile robots*. We propose a learning technique that combines incremental learning in constructive neural networks with the computational efficiency of instance-based learning and dynamic programming methods. Moreover, we present a robot control architecture that is suitable for reinforcement learning. For state estimation, we use a position tracking system which is based on sonar sensor readings. The learning algorithm is implemented as a module of the control architecture. In selected experiments, we show that the proposed learning algorithm allows the transfer of information between related reinforcement learning tasks in a real robot.

We do not attempt to address in this paper all the issues related to continual learning. For example, we do not discuss methods for exploration. Exploration is surely needed in changing environments. We present preliminary results. The learning algorithm does not include any means of forgetting yet. In the presentation, we focus on a description of the algorithm, but do not give a detailed quantitative comparison of the learning methods.

The paper is organised as follows. In Section 2, we introduce reinforcement learning. We provide a formal framework for learning from reinforcement and summarise methods for finding an optimal policy. In Section 3, we critically examine previous work on learning with adaptive state spaces, introduce a criterion for the distinction of states, and discuss the main challenges of incremental learning. In Section 4, we introduce a novel constructive neural-network model, called dynamic cell structures, which can be used for multi-task reinforcement learning. In Section 5, we describe the control architecture of the robot, which consists of modules for low-level control, vision-based object detection, position tracking using sonar sensors, and learning. The proposed learning algorithm is applied to a mobile robot which has to learn a navigation task in different environments. Preliminary results of this application are given in Section 6. Finally, we draw some conclusions in Section 7.

2   Reinforcement Learning

2.1   Framework

This section provides a formal notation for describing the environment and the agent. It has been adapted from previous work by Kaelbling *et al.* [9]. We approximate the world by making the state space, the action space and the time discrete. At each time step, an action is executed, and the world instantaneously transforms itself from its previous state to the new state that results from that action. These assumptions are universally considered valid for real robots.

The agent's environment can be defined as tuple $(\mathcal{X}, \mathcal{A}, T)$, where

- $\mathcal{X}$ is a finite set of states of the world,
- $\mathcal{A}$ is a finite set of actions, and
- $T : \mathcal{X} \times \mathcal{A} \rightarrow \mathcal{X}$ is the state-transition function, giving for each world state and agent action a probability distribution over world states.

The agent's reward is defined by the reward function, $R : \mathcal{X} \times \mathcal{A} \rightarrow \mathbb{R}$, giving the expected immediate reward gained by the agent for taking each action in each state. A policy π is a mapping from $\mathcal{X}$ to $\mathcal{A}$ specifying an action to be taken in each situation.

The tuple $(\mathcal{X}, \mathcal{A}, T, R)$ defines a Markov decision process (MDP) if future states and rewards are conditionally independent from past states and actions. The state and reward at time $t + 1$ is dependent only on the state at time t and the action at time t. This is known as the Markov property.

In the description of the agent, we have to take into account that the agent may not be able to determine the state it is currently in with complete reliability, as it is the case for a robotic agent. We define the agent's perception of the environment as a tuple (Ω, O), where

- Ω is a finite set of observations the agent can experience of its world, and
- $O : \mathcal{X} \times \mathcal{A} \rightarrow \Pi(\Omega)$ is the observation function, which gives for each action and resulting state a probability distribution over possible observations.

The tuple $(\mathcal{X}, \mathcal{A}, T, R, \Omega, O)$ defines a partially observable Markov decision process (POMDP). Kaelbling *et al.* [9] proposed to decompose the problem of controlling a POMDP into two parts, as shown in Figure 1. The agent makes observations, $o \in O$, and generates actions, $a \in \mathcal{A}$. It keeps an internal belief state, b, that summarises its previous experience. A belief state is a discrete probability distribution over the set of world states, $\mathcal{X}$, representing for each state the agent's belief that it is currently occupying that state. The state estimator (SE) is responsible for updating the belief state based on the last action, the current observation, and the previous belief state. The component π is the policy. As before, it is responsible for generating actions, but this time as a function of the agent's belief state rather than the state of the world.

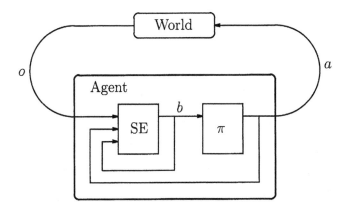

Fig. 1. A POMDP agent can be decomposed into a state estimator SE and a policy π.

2.2 State Estimation

When the agent is provided with a POMDP model of its environment and the agent uses the belief state as the state space on which to specify its policy, the key difficulty is that this state space is continuous and has as many dimensions as there are world states. This makes the task of finding an optimal POMDP control strategy computationally intractable. On the other hand, we have efficient computational techniques for solving (completely observable) Markov decision processes. That is, a POMDP problem can be solved if the agent can be provided with a state estimator for resolving the uncertainty about the state. In many domains, such state estimation methods are indeed available. For example, Kalman filters [15] and Bayesian methods [5, 8] have been used for state estimation in mobile robot navigation. The belief state has been represented by unimodal Gaussian distributions (Kalman filter), sets of Gaussians, probability grids, or sampling methods (Markov localisation).

A position tracking system which is based on sonar-sensor readings and that can be used as state-estimator in robot-navigation tasks is described in Section 5. So, in the following, we assume the availability of a state estimator. The output of the state estimator will be the most likely state, i.e., the world state with the highest probability. Moreover, we assume that $\mathcal{S}$ is a finite set of discrete states of the agent, and there is a mapping $\mathcal{X} \to \mathcal{S}$ so that $(\mathcal{S}, \mathcal{A}, T, R)$ is a finite state Markov decision process. That is, there is no hidden state. The information about the agent's state, $s \in \mathcal{S}$, combined with the POMDP model of the environment is sufficient to predict, in principle, the future interaction of the agent with the environment.

2.3 Finding a Policy Given a Model

Given a policy π and a reward function R, the value of a state $s \in \mathcal{S}$ is the expected value of sums of the rewards to be received at each future time step, discounted by how far into the future they occur. That is,

$$V_\pi(s) = \mathrm{E}\left[\sum_{t=0}^{\infty} \gamma^t r_t\right]$$

where r_t is the reward received on step t of executing policy π after starting in state s. A closely related quantity $Q_\pi(s, a)$ is the value of state s when action a is executed first and policy π is followed thereafter. The discount factor, $0 \le \gamma < 1$, controls the influence of the rewards in the distant future. When $\gamma = 0$, the value of a state is determined entirely by rewards received at the next step. We are generally interested in problems with further horizon and so we set γ to be near 1.

So far, we have considered learning tasks in which the agent-environment interaction goes on continually without limit. The above definition of $V_\pi(s)$ can be applied to episodic tasks (trials) as well. Episode termination corresponds to entering a special *absorbing state* that transitions only to itself and that generates only rewards of zero.

The function V can be defined recursively [29] as:

$$V_\pi(s) = R(s, a) + \gamma \sum_{s' \in \mathcal{S}} T(s, a, s') V_\pi(s').$$

We are interested in finding an optimal policy, i.e., a policy π that maximises $V_\pi(s)$ for all $s \in \mathcal{S}$. We shall use π^* to refer to an optimal policy for an MDP, and express the optimal value and Q functions as $V(s)$ and $Q(s, a)$. The optimal value function is unique and can be defined as the solution to the simultaneous equations

$$V(s) = \max_{a \in \mathcal{A}}\left(R(s, a) + \gamma \sum_{s' \in \mathcal{S}} T(s, a, s') V(s')\right)$$

for all $s \in \mathcal{S}$. Given the optimal value function, the optimal policy can be specified as:

$$\pi(s) = \arg \max_{a \in \mathcal{A}}\left(R(s, a) + \gamma \sum_{s' \in \mathcal{S}} T(s, a, s') V(s')\right).$$

There are many methods for finding optimal policies for MDPs. One way is to find the optimal value function. It can be determined by a simple iterative algorithm called *value iteration* [29]. The algorithm is shown in Figure 2. It stores approximations to $V(s)$ and $Q(s, a)$. The algorithm terminates when the maximum difference between two successive value functions is less than some ϵ.

2.4   Learning an Optimal Policy

In the previous section, we have assumed that we had already a model. The model consists of knowledge of the state transition-probability function T and

$V_1(s) \leftarrow 0$ for all $s \in \mathcal{S}$
$t \leftarrow 1$
loop
    $t \leftarrow t + 1$
    loop for all $s \in \mathcal{S}$
        loop for all $a \in \mathcal{A}$
$$Q_t(s, a) \leftarrow R(s, a) + \gamma \sum_{s' \in \mathcal{S}} T(s, a, s') V_{t-1}(s')$$
        end loop
$$V_t(s) \leftarrow \max_{a \in \mathcal{A}} Q_t(s, a)$$
    end loop
until $|V_t(s) - V_{t-1}(s)| < \epsilon$ for all $s \in \mathcal{S}$

Fig. 2. Finding the optimal value function through value iteration.

the reinforcement function R. We are interested in applications for which such a model is not known in advance. The learning agent must interact with its environment directly to obtain information which can be processed to produce an optimal policy.

There are two classes of reinforcement learning algorithms: *model-free* algorithms which learn a policy without learning a model, e.g., Q-learning [32] and the TD(λ) algorithm [28], and *model-based* algorithms which learn such a model and use it to find an optimal policy, e.g., prioritised sweeping [24] and real-time dynamic programming [1]. For the general case, it is not clear yet which approach is better in which circumstances.

3   Learning with Adaptive State Spaces

In the ideal case, the agent's state-space is the set of all possible partial solutions to the given learning task. Actions are ways of getting from one state to another. A successful solution consists of a sequence of actions in that state space. In most RL applications, the agent designer predefines the possible perceptual inputs, the internal state-space, and the actions of the learning agent. Pre-selecting the state space by a human programmer may be very effective because it can provide a useful bias for the learning algorithm. However, this approach will fail if applied to multi-task learning. Generally, the learning agent does not know beforehand about the perceptual input it will encounter. So, the agent should be able to select its state space automatically while learning.

3.1   Adaptive Resolution Methods

The automatic extraction of relevant states is particularly important for learning in multidimensional continuous state spaces. What we would like to do is partition the environment into regions of states that can be considered the same for the purpose of learning and generating actions. If this discretisation is too coarse-grained, the agent will fail. If the discretisation is too fine-grained the number of states is too big more than needed and learning becomes computationally intractable. To be able to generalise efficiently over input, we need different granularity in different regions of state space. This problem has been addressed in methods that use adaptive resolution such as variable resolution dynamic programming (VRDP) [21] and parti-game [22].

The VRDP algorithm uses a *kd*-tree, which is similar to a decision tree, to partition the state space into coarse regions. The coarse regions are refined into detailed regions, but only in parts of the state space which are predicted to be important. This notion of importance is obtained by running 'mental trajectories' through state space, i.e., executing the current optimal actions off-line. It is assumed that those states encountered during mental practice are states which are particularly important to know about it detail. A disadvantage of VRDP is that it partitions the state space into high resolution everywhere the agent visits, potentially making many irrelevant distinctions. An advantage of VRDP is that the part of the problem which truly requires learning (building the model of the environment) is entirely decoupled from the part which requires computation (the search for an optimal strategy). Dynamic programming is performed only on the extracted states.

Parti-game is another adaptive resolution method. It addresses the problem of learning to achieve goal configurations in deterministic high-dimensional continuous spaces by dynamically partitioning the space into hyper-rectangular cells of varying sizes, represented using a *kd*-tree data-structure. Parti-game takes advantage of memorised instances in order to speed up learning. The results reported by Moore [20] indicate that instance-based methods are well-suited to problems in which the state-space granularity is changing.

3.2   Memorising Learning Experience

Instance-based learning methods such as parti-game memorise as much learning experience as possible. In the robotics domain, this is certainly profitable. Agents should learn in as few trials as possible, and learning experience is usually expensive. In robotics, gathering experience can take several orders of magnitude more time than the necessary computation. Learning two things at the same time – evolving the state space as well as trying to find an optimal policy – does not necessarily take more time than learning the policy on its own for a given state space. If the agent is able to learn a good representation of the state space, it will generalise well, which in turn will speed up the learning process. However, this is only possible if the agent makes efficient use of the experience. If the learner incorporates experience merely by averaging in its current, flawed state space granularity, it is bound to attribute experience to the wrong states.

In the learning methods above, we can distinguish between representational tools and learning paradigms [23]. The representational tools in VRDP and parti-game are *kd*-trees. They are used to partition the state space, approximate the value function, and store learning instances. The learning paradigm defines what the representation is used for, how training data is used to modify the representation, whether exploratory actions are performed, and other related issues. It also specifies a criterion for splitting (and possibly merging) states. Recently, McCallum [18] defined such a criterion. He suggested to distinguish states that have different policy actions or different utilities, and merge states that have the same policy action and same utility. This *utile distinction* test yields a task-dependent state representation.

3.3   Utility-Based State Distinction

Often a perceptually aliased state that affects utility will have wildly fluctuating reward values. However, we cannot base a state splitting test solely on reward variance because some changes in reward are caused by the stochastic nature of the world. The agent must be able to distinguish between fluctuations in reward caused by a stochastic world, and fluctuations that could be better predicted after adding a state distinction. As suggested by McCallum [18], the agent can use statistical tests based on future discounted reward to find state distinctions that help predict reward.

A state model, which has been created by state distinctions, is said to be Markov with respect to reward if and only if future reward is conditionally independent of past states and actions, given the current state and action [18]. That is, $P(r_{t+1}|s_t, a_t, s_{t-1}, a_{t-1}, \ldots, s_0) = P(r_{t+1}|s_t, a_t)$ where r_{t+1} is the expected future discounted reward at time $t+1$. A state is Markov with respect to reward if knowledge of past states does not help predict future reward from that state. McCallum proposed to keep reward statistics associated with incoming transitions to a state. If the state satisfies this Markov property, its separated reward statistics are expected to be the same. If, on the other hand, there is a statistically significant difference in future discounted reward between the statistics depending on where the agent came from, then knowledge of which state the agent came from does help predict reward. In this case, the state should be split.

It is still an open question which non-parametric statistical test should be used. McCallum [18] has reported promising results using the Kolmogorov-Smirnov test, which asks whether two distributions came from the same source.

3.4   Incremental Learning

Generally, changes in the environment occur unpredictably. They could not in principle have been modelled, nor could their occurrence have even been announced. Adjusting to this sort of event is an important problem for any learning agent [20]. Being confronted with changes in the environment, the system may forget what it has previously learnt. This problem is usually referred to as stability-plasticity dilemma – how can a learning agent preserve what it has previously learnt, while continuing to incorporate new knowledge.

3.5 Discussion

At this point, we want to summarise the lessons learnt from previous work on learning with adaptive state spaces. First, tree-structured representations such as kd-trees have been used successfully to find a task-dependent segmentation of the agent-internal state space. As representational tools, they support incremental development. However, tree-structured representations are rather impractical when they are used for continual learning, which does include the need for forgetting. Second, for learning with adaptive state spaces in general and for robots in particular, it is essential to make efficient use of experiences encountered in previous learning trials. To be able to access the past experiences efficiently, they should be stored locally in the state-space partitions. Third, McCallum's utile distinction test provides us with a criterion for deciding whether a particular state distinction helps in predicting reward. It has been used in learning algorithms to find distinctions in a discrete sensory space [18].

We propose to use constructive neural networks as representational tools for continual learning. Ontogenic neural networks that perform a segmentation of the input space by unsupervised learning seem to fulfil the requirements of continual learning better than kd-trees. The main issue in constructive techniques is to decide when and where to add new neurons. We suggest to use McCallum's utile distinction test as selection criterion. It is hoped that the utile distinction test will perform equally well for distinctions in continuous space.

4 Multi-task Learning with Dynamic Cell Structures

The continual-learning method proposed in this paper performs two kinds of learning: instance-based population of the state space with raw experience and dynamic programming to calculate expected future reward. Both components are well separated. The instance-based part is used to extract the relevant state space, then dynamic programming is performed only on the extracted states. This approach was motivated by McCallum's U-Tree algorithm [18]. In fact, we use McCallum's learning paradigm together with a representational tool – a constructive neural network – that supports incremental and continual learning.

4.1 Representation

By value iteration, we compute payoff predictions over a discrete set of agent-internal states. Given that the perceptual space is usually multi-dimensional and continuous, there is the need to identify the individual states and to approximate their values. For multi-task learning, it should be possible to add input patterns in a remote area of the input space incrementally, without affecting the already learnt input/output relations. Function approximators based on local representations can do so. On the other hand, they perform only local generalisation which often prevents them from discovering global relations in the input space, which function approximators based on distributed representations are capable of.

In this work, we use ontogenic neural networks as function approximators. Unsupervised ontogenic neural networks have been applied successfully to vector quantisation, data visualisation, and clustering [7]. Fritzke [6] proposed an incremental self-organising network model, which is called growing cell structures. This model is able to generate dimensionality reducing mappings. These mappings are able to preserve neighbourhood relations in the input data and have the property to represent regions of high input density on correspondingly large parts of the topological structure. In contrast to Kohonen's self-organising feature map [12], which serves similar purposes, neither the number of units nor the exact topology has to be predefined in this model. Instead, a growth process successively inserts units and connections. In the following, we introduce the dynamic cell structures network model, which has been adapted from Fritzke's growing cell structures [6].

The model comprises a set C of nodes. Each node $c \in C$ has an associated n-dimensional reference vector, w_c, which indicates the centre of its receptive field in the input space $\mathbb{R}^n$. A given set of nodes with their reference vectors defines a particular partition of the input space. The receptive field of a cell c consists of those points in $\mathbb{R}^n$ for which w_c is the nearest of all currently existing reference vectors. The cells in the network represent the agent-internal states of the reinforcement-learning agent. Between certain pairs of nodes there are edges indicating neighbourhood. The resulting topology is strictly k-dimensional whereby k is some positive integer chosen in advance with $k \leq n$. The basic building block and also the initial configuration of each network is a k-dimensional simplex, consisting of $k + 1$ fully connected nodes. For $k = 2$ this is a triangle, for $k = 3$ a tetrahedron, and for $k > 3$ the structure is referred to as hypertetrahedron. See Figure 3.

4.2   Algorithm

We propose an algorithm, referred to as dynamic cell structures, that performs a dynamic segmentation of the input states and that keeps track of past learning experiences. The partitions of the input space correspond to agent-internal

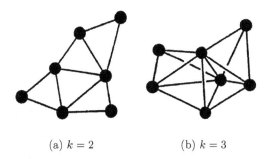

(a) $k = 2$ (b) $k = 3$

Fig. 3. Fritzke's cell structures of different dimensionality k.

states, and the learning experiences are linked to the individual partitions in which they have occurred.

Learning begins by finding a suitable initial segmentation of the input space. In the pre-learning phase, the agent collects initial experience by executing random actions or actions provided by a reinforcement program that acts as a demonstrator. The learning experience is stored as a chain of transition instances, $T_t = \langle T_{t-1}, a_{t-1}, o_t, r_t \rangle$, where the encountered observations are then used as training instances for finding the initial network by unsupervised learning as follows:

(1) Choose an observation o_t from the set of all transition instances and present it to the network.
(2) Determine the best-matching unit c_b (the unit with the nearest reference vector): $\|w_b - o_t\| \leq \|w_i - o_t\|$ for all $c_i \in \mathcal{C}$.
(3) Move c_b and its direct topological neighbours (those unit which are connected to c_b by an edge) towards o_t by fractions ϵ_b and ϵ_n, respectively, of the total distance: $w_b = \epsilon_b(o_t - w_b)$ for the best matching unit u_b and $w_i = \epsilon_n(o_t - w_i)$ for all topological neighbours c_i of c_b.

The action a_{t-1} in T_t was selected on the basis of observation o_{t-1}, which is part of T_{t-1}. Therefore, the algorithm associates each transition instance T_t with the best-matching node for the observation o_{t-1} of the predecessor T_{t-1}.

The learning phase consists of the following steps:

(1) The agent makes a step in the environment. It records the transition as an instance, and puts it to the end of the chain of instances. The algorithm stores the instance T_t with the best-matching node for the observation o_{t-1} of the predecessor T_{t-1}.
(2) For each step in the world, the agent does one sweep of value iteration, with the nodes of the network as states:

$$Q(s, a) \leftarrow R(s, a) + \gamma \sum_{s'} T(s, a, s') V(s')$$

with $V(s') = \max_{a \in \mathcal{A}} Q(s', a)$. Both $R(s, a)$ and $T(s, a, s')$ can be calculated directly from the recorded instances.
(3) After every l steps, the agent tests whether newly added information or dynamic programming (value iteration) has changed the transition utility statistics enough to warrant adding any new distinctions to the agent's internal state space. In this case, the agent selects a parent node, creates a fringe node in the direct neighbourhood of the parent, and uses the Kolmogorov-Smirnov test to compare the distributions of future discounted reward associated with the same action from the fringe node and any other node. If the test indicates that all the distributions have a statistically significant difference, this implies that promoting the relevant fringe node into a non-fringe node will help the agent to predict reward and so the change is retained.

The distribution of the expected future discounted reward associated with a particular node is composed of the set of expected future discounted reward values associated with the individual instances in that node. The expected future discounted reward of instance T_i, $Q(T_i)$, is defined as $Q(T_i) = r_i + \gamma \, V \, (C(T_{i+1}))$. The node to which instance T belongs to is termed $C(T)$.

As parent nodes, we select nodes that have a large deviation between their instances' utility values. There are many possible variations on techniques for selecting the parent node. A straightforward one is to sort the nodes in descending order according to the deviation between their instances' Q-values and to select the nodes in this order.

The reference vector of the fringe node is found by unsupervised learning. As training instances we use only the observations associated with the selected parent node. During the unsupervised learning, we allow only the reference vectors of the parent and fringe node to change. The reference vectors of all other units are not changed in this process.

The learning algorithm attempts to solve two problems at the same time – evolving the state space by adding fringe nodes to the neural network and finding an optimal policy by value iteration on the evolved state space. An example is given in the Section 5 and 6.

5   A Control-Architecture for the Pioneer 1 Mobile Robot

Controlling a mobile robot in the real world generally involves complex operations for motor control, sensing, and high-level control. In order to deal with this complexity, robot designers often use hierarchical, behaviour-based control architectures. That is, the complex behaviour of the robot visible in the real world results from the interaction and coordination of a set of basic behaviours. We have used reinforcement learning to find individual basic behaviours. In the following, we will describe the system architecture we have used on a Pioneer 1 mobile robot which integrates behaviour-based control, sensing, and learning.

5.1   Control, Perception, and Learning

The Pioneer 1 is powered by two DC motors. It can move with an approximate maximum speed of 60 cm per second. The robot is equipped with seven ultrasonic proximity sensors – one on each side and five forward facing – and a colour CCD camera, which is mounted in a fixed position on top of the robot's console. The Pioneer used in the experiments is also equipped with a 2 axis, 1 degree-of-freedom gripper. The robot control software runs off-board, on a remote UNIX workstation, with which the robot communicates through a radio link. The remote software receives status updates from the robot and its sonar sensors at a frequency of 10 Hz. The maximum command rate for changing the motion direction is only 2 Hz (or less) due to limitations of the on-board controller. The video images are also processed off-board. The maximum frame rate depends on the image processing operations performed. Object detection

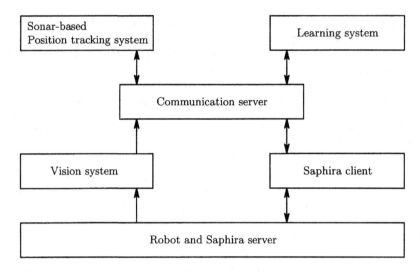

Fig. 4. The system architecture of the robot control software. Perception modules are on the left, control modules on the right.

using colour segmentation is done in real-time at a rate of about two frames per second.

Figure 4 shows the basic architecture of the robot control software. It consists of several modules, which are implemented as different processes communicating with each other through a communication server. At the lowest level of the system architecture, we find the Saphira server [13], which runs on board the robot. The function of the server is to control the drive motors and the gripper and to collect data from the position encoders of the wheels and the sonar sensors. The server receives target values of the wheel velocities from the Saphira client running on the host computer. The Saphira client integrates a number of functions such as sending commands to the server, gathering information from the robot's sensors, and micro-tasking.

The robot perceives the environment through the colour camera and the sonar sensors. The vision system performs colour segmentation. It can detect objects of three distinct colours: red and blue soda cans and the green labels of the collection points. The position tracking system computes the robot's position and orientation in the environment from previous distance readings.

The learning system is the part of the system architecture that is responsible for adaptability and learning. Its task is to find action policies through reinforcement learning. As described in Section 2, a POMDP agent can be decomposed into a state estimator and a policy. In the architecture given above, the agent's observations are the readings from the sonar sensors and wheel encoders, the state estimator is the position tracking system, the agent's belief is the position and orientation in the environment, and the policy is computed by the learning system.

5.2   Behaviours and Action Selection

Learning a complex task from scratch is currently impractical for mobile robots. However, it is often possible to learn the task in stages (incremental learning) or to learn some part of the task. For example, a can-collecting robot could learn the navigation from reinforcement while all other task-achieving modules are coded by the robot designer. That is, we could solve a complex task by integrating autonomous learning and manual programming.

The output of the learning system are actions which are to be executed by the robot's actuators. In many cases where reinforcement learning is used on a real robot, we will need some safety mechanism. For example, while learning to navigate an environment, we need to ensure that the robot does not drive with full speed against a wall.

Behaviour-based, hierarchical control mechanisms such as the Saphira architecture by Saffioti *et al.* [26] fulfil the requirements mentioned above. The idea is to decompose the control problem into small units of control. These basic behaviours are often relatively simple and can be added incrementally. Strictly speaking, a behaviour is a mapping from the internal state to control actions of a robot within a restricted context. Moreover, the behaviours have hierarchy levels assigned. An action issued by an individual behaviour will be executed if the behaviour is active and the action is not over-ruled by a behaviour higher up in the hierarchy. Actions issues by behaviours of the same hierarchy level are merged.

The task of collecting cans can be decomposed as follows. The basic behaviour of the highest hierarchy level is Safe_Velocity_Ctrl which reduces the wheel velocity if the distance readings of the front sonars get below a certain level. There are two other basic behaviours, Navigation and Gripper_Ctrl. The former will turn the robot towards a can to be collected or the collection point where the can is to be dropped, while avoiding to bump into walls. The latter will stop the robot and open or close the gripper at the right time. In the section on experiments, we describe how the Navigation behaviour can be learnt.

5.3   Localisation and Position Tracking

Most existing localisation methods for mobile robots make simplifying assumptions about the properties of the sensors. These methods therefore work well only when the inherent assumptions hold for the particular robot, its behaviour, and its environment. Many methods, for example, assume a large number of evenly spaced sensors, which render them useless in robots with very few sensors.

In comparison to mobile robots that have a ring of sonars, the sensing capabilities of the Pioneer 1 are rather limited. At first, we tried to use a Markov localisation method [2]. However, this approach failed. The robot became lost when the sonar sensor readings were sparse and noisy, for example, when the robot was moving diagonally through a corridor. In this situation, the walls of the corridor reflect the sonar beams and hardly any distance readings from the front sonars are correct.

In further investigations, we developed a localisation method [8] that works reliably in our setup. The approach consists of three steps. First, we compute a two-dimensional feature space by applying a straight-line Hough transform [14] to the sonar readings. Second, we perform template matching in the feature space by using the world map as reference pattern. Third, we use the correlation counts obtained in the previous step to update a position probability grid.

This method, which is to our knowledge the only sonar-based position tracking system for the Pioneer 1 to date, is less dependent on individual sonar sensor readings than a traditional Markov localisation approach. Rather than using the product of the likelihoods, $p(s|L)$, the likelihood of observing the sensor reading s at the position L, we use the correlation count (a combined feature) for the computation of position probabilities. In addition, we use the detection of wall-like segments in the sonar data to detect situations in which the the current sensor information is insufficient for localisation. In those situation, the sensor information is not used to update the position probabilities. The method integrates a feature-based detection method with a dense-sensor matching technique by using the Hough transform for feature detection and a grid-based approach to update a distribution of position probabilities.

6   Experiments on Continual Learning

We have applied the RL algorithm of Section 4 to a mobile robot learning a navigation task in different environments. The control architecture of the robot is described in the previous section. We have chosen a scenario in which the robot's task is to reach a sequence of goal positions, which have been chosen at random. The learner receives reward when it finds the goal and punishment when it bumps into a wall. The algorithm should be evaluated with respect to its ability to segment the agent-internal state space and to adapt state space and policy, for example, when some features of the environment change.

6.1   Learning Scenario

The Pioneer 1 robot was operated in a laboratory environment. As shown in Figure 5, the robot arena consists of straight wall segments. For the robot, it is difficult to navigate due to the existence of many corners and edges.

The input to the learning system, corresponding to the agent-internal belief state b, at each time step is the vector $\langle x, y, \theta, \alpha \rangle$, where $\langle x, y, \theta \rangle$ denotes the estimated position and orientation of the robot in the environment, which are provided by the position tracking system [8], and $-\pi \leq \alpha \leq \pi$ is the angle to the goal with respect to the robot's current orientation. Values $\alpha < 0$ denote that the goal is on the left-hand side and $\alpha > 0$ on the right-hand side of the robot.

The output of the learning system are the three possible actions: turn left, move straight-on, and turn right. Each command is executed by the robot for a period of 0.1 seconds. It may take the robot up to 10 seconds to reach the goal position or to bump into a wall. The robot is said to reach the goal if it

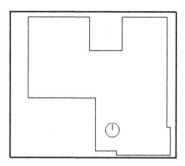

Fig. 5. The Pioneer 1 mobile robot and its test environment. The size of the robot arena is 5.2 m × 4.4 m.

comes within a radius of 40 cm to that position. In this case, the learner receives a reinforcement of $r = 1$. If the robot bumps into a wall, the performance feedback is $r = 0$, independent of the distance to the goal at the time of the collision. A learning trial is finished if the robot reaches the goal or bumps into a wall. In either case, a new goal position is chosen at random.

The navigation behaviour which is learnt from reinforcement is later used for the can-collection task, in which the the angle α corresponds to the visual angle the object to be picked up is in, that is provided by the vision system.

Learning a suitable policy for the navigation task is difficult. First, the interaction with the environment is stochastic. The position information is subject to an error of up to 30 cm in xy-direction and 10 degree in θ-direction in the worst case. Given the same input to the learning system, in one situation there might be enough space for a full turn in front of an obstacle and in another situation there might not. That is, the performance feedback R is drawn from a probability distribution. Second, the feedback is extremely delayed, considering that an action sequence can consists of up to 100 steps. To speed up the learning process, we use a reinforcement program that acts as an demonstrator in the early phase of the learning process. In fact, this is a hand-written controller, turning the robot to the right if the goal is on the right-hand side, turning it to the left if the goal is on the left, and moving it straight-on otherwise. Obviously, this behaviour is sub-optimal in the presence of edges in the environment. However, it makes sure that the learner receives reward from time to time in the early phase of the learning.

6.2 Experimental Evaluation

The agent-internal state space in the navigation task has $k = 4$ dimensions. Since the visualisation of such a space is difficult, we have used dynamic quantisation

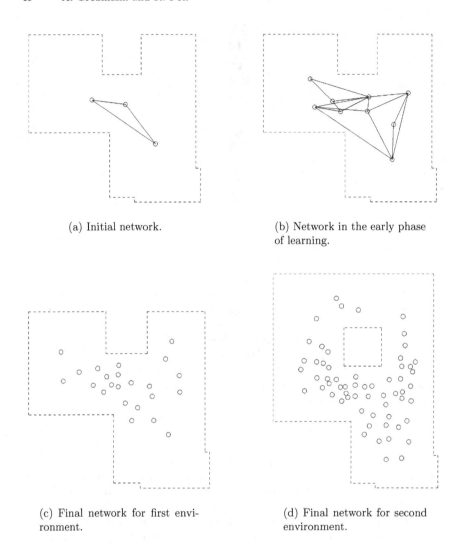

(a) Initial network.

(b) Network in the early phase of learning.

(c) Final network for first environment.

(d) Final network for second environment.

Fig. 6. Dynamic cell structures are used to segment the $\langle x, y \rangle$-plane for the robot navigation task. In (c) and (d), the edges between the nodes are omitted.

only in the xy-plane. The segmentation for θ and α was done manually. We have used 6 intervals in θ and 4 intervals in α direction. See Figure 6 for an example. The learner starts with a small number of nodes – see Figure 6(a) (for $k = 2$ input dimensions, this is 3 nodes corresponding to 3 agent-internal states) – and then divides the state space further by adding nodes while performing dynamic programming. If the learning task does not change, we are going to reach a situation in which adding nodes does not help in predicting reward. That is, the network does not change anymore, see Figure 6(c). After this task has been learnt, the environment is extended, which requires a modification of the state-space and action policy. Taking network (c) as initial network, the result of the adaptation is shown in Figure 6(d).

The experiments performed so far on learning the task from scratch indicate a speed-up of about 20 percent due to better generalisation in comparison to learning the same task using a predefined state space (160 xy-tiles of 30 cm length, 6 θ-intervals, 3 α-intervals). The corresponding neural network consists of about 25 nodes in the xy-plane.

The speed-up in multi-task learning that can be achieved depends on the reinforcement learning method used. There are two classes of RL algorithms: *model-free* algorithms which learn a policy without learning a model, e.g., Q-learning and the TD(λ) algorithm, and *model-based* algorithms which learn such a model and use it to find an optimal policy, e.g., prioritised sweeping and real-time dynamic programming [10]. A continual-learning agent has typically to learn simple tasks first before progressing to more complex ones. This approach of 'shaping' requires constant modifications of the reward landscape, which means that an agent implemented with Q-learning will be re-learning most or all of its Q-values again and again. With respect to this changing reward problem, model-based methods have an advantage because the agent could transfer its state-transition mapping to a similar domain or task easily. We are currently performing a detailed performance evaluation of model-based (value iteration, prioritised sweeping) and model-free learning paradigms (Q(λ)-learning) in connection with the use of dynamic cell structures as representational tools.

7   Conclusions

We have argued that there is a close connection between adaptive state-space quantisation and continual learning. This was the motivation for developing a novel method for learning multiple tasks from reinforcement. This technique, which has been applied to a mobile robot, performs a quantisation of the agent-internal state space using constructive neural networks. Currently, we are in the process of evaluating the effectiveness of the method for multi-task learning. Preliminary experiments have shown that the method can be used to control a Pioneer 1 mobile robot collecting soda cans. The method enables the robot learner to re-use the policy and model of the environment learnt so far when the environment changes.

The experimental results indicate that the utile distinction test can be applied successfully to find distinctions in a continuous sensory space if unsupervised-learning constructive neural networks are used as representational tools. To deal with the uncertainty in the robot's sensations and actions, we have used a sonar sensor-based position tracking system as state estimator. Learning the navigation task from the raw sonar sensor readings only would not have been possible.

References

1. A. G. Barto, S. J. Bradtke, and S. P. Singh. Learning to act using real-time dynamic programming. *Artificial Intelligence*, 72(1):81–138, 1995.
2. W. Burgard, D. Fox, and D. Hennig. Fast grid-based position tracking for mobile robots. In G. Brewka, C. Habel, and B. Nebel, editors, *Proceedings of the 21st Annual German Conference on Artificial Intelligence (KI-97): Advances in Artificial Intelligence*, pages 289–300. Springer, 1997.
3. R. A. Caruana. Learning many related tasks at the same time with backpropagation. In G. Tesauro, D. Touretzky, and T. Leen, editors, *Advances in Neural Information Processing Systems*, volume 7, pages 657–664. MIT Press, 1995.
4. R. A. Caruana. Algorithms and applications for multitask learning. In *Proceeding of the Thirteenth International Conference on Machine Learning (ICML-96)*, pages 87–95, Bari, Italy, 1996. Morgan Kaufmann.
5. D. Fox, W. Burgard, F. Dellaert, and S. Thrun. Monte Carlo localisation: Efficient position estimation for mobile robots. In *Proceedings of the Sixteenth National Conference on Artificial Intelligence (AAAI-99)*, 1999.
6. B. Fritzke. Growing cell structures – a self-organizing network for unsupervised and supervised learning. Technical Report TR-93-026, International Computer Science Institute, Berkeley, CA, USA, 1993.
7. B. Fritzke. Unsupervised ontogenic networks. In E. Fiesler and R. Beale, editors, *Handbook of Neural Computation*, chapter C 2.4. Institute of Physics and Oxford University Press, 1997.
8. A. Großmann and R. Poli. Robust mobile robot localisation from sparse and noisy proximity readings. In *Proceedings of the IJCAI-99 Workshop on Reasoning with Uncertainty in Robot Navigation (RUR-99)*, 1999.
9. L. P. Kaelbling, M. L. Littman, and A. R. Cassandra. Planning and acting in partially observable stochastic domains. *Artificial Intelligence*, 101:99–134, 1998.
10. L. P. Kaelbling, M. L. Littman, and A. W. Moore. Reinforcement learning: A survey. *Artificial Intelligence Research*, 4:237–285, 1996.
11. Z. Kalmár, C. Szepesvári, and A. Lőrincz. Module based reinforcement learning for a real robot. Presented at the Sixth European Workshop on Learning Robots, Brighton (EWLR-6), UK, 1997.
12. T. Kohonen. Self-organising formation of topologically correct feature maps. *Biological Cybernetics*, 43(1):59–69, 1982.
13. K. Konolige. *Saphira Software Manual. Version 6.1*. ActivMedia, Inc., Peterborough, NH, USA, 1997.
14. V. F. Leavers. *Shape Detection in Computer Vision Using the Hough Transform*. Springer, London, UK, 1992.
15. J. K. Leonard and H. F. Durrant-Whyte. *Directed Sonar Sensing for Mobile Robot Navigation*. Kluwer Academic, 1992.
16. S. Mahadevan and J. Connell. Automatic programming of behaviour-based robots using reinforcement learning. *Artificial Intelligence*, 55(2-3):311–365, 1992.
17. M. J. Matarić. Reward functions for accelerated learning. In *Proceedings of the Eleventh International Conference on Machine Learning*, pages 181–189. Morgan Kaufmann, 1994.
18. A. K. McCallum. *Reinforcement learning with selective perception and hidden state*. PhD thesis, Department of Computer Science, University of Rochester, Rochester, NY, USA, 1995.

19. L. A. Meeden. An incremental approach to developing intelligent neural network controllers for robots. *IEEE Transactions on Systems, Man, and Cybernetics*, 26, 1996. Special Issue on Learning Autonomous Robots.

20. A. W. Moore. *Efficient memory-based learning for robot control.* PhD thesis, Computer Laboratory, University of Cambridge, Cambridge, UK, 1990.

21. A. W. Moore. Variable Resolution Dynamic Programming: Efficiently learning action maps in multivariate real-valued state spaces. In L. A. Birnbaum and G. C. Collins, editors, *Machine Learning: Proceedings of the Eighth International Workshop (ML-91)*. Morgan Kaufmann, 1991.

22. A. W. Moore. The parti-game algorithm for variable resolution reinforcement learning in multi-dimensional state-spaces. In J. D. Cowan, G. Tesauro, and J. Alspector, editors, *Advances in Neural Information Processing Systems*, volume 6, pages 711–718. Morgan Kaufmann, 1994.

23. A. W. Moore, C. Atkeson, and S. Schaal. Memory-based learning for control. Technical Report CMU-RI-TR-95-18, Robotics Institute, Carnegie Mellon University, Pittsburgh, PA, USA, 1995.

24. A. W. Moore and C. G. Atkeson. Prioritised sweeping: Reinforcement learning with less data and less real time. *Machine Learning*, 13, 1993.

25. M. B. Ring. *Continual learning in reinforcement environments.* PhD thesis, University of Texas, Austin, TX, USA. Available via the URL http://www-set.gmd.de/~ring/Diss/, 1994.

26. A. Saffioti, E. Ruspini, and K. Konolige. Blending reactivity and goal-directedness in a fuzzy controller. In *Proceedings of the IEEE International Conference on Fuzzy Systems*, pages 134–139, 1993.

27. S. Schaal and C. G. Atkeson. Robot juggling: An implementation of memory-based learning. *Control Systems Magazine*, 14, 1994.

28. R. S. Sutton. Learning to predict by the methods of temporal differences. *Machine Learning*, 3:9–44, 1988.

29. R. S. Sutton and A. G. Barto. *Reinforcement learning: An Introduction.* MIT Press, 1998.

30. S. Thrun. *Explanation-based neural network learning: A lifelong learning approach.* Kluwer Academic, Norwell, MA, USA, 1996.

31. S. Thrun and L. Pratt, editors. *Learning to Learn.* Kluwer Academic, Norwell, MA, USA, 1998.

32. C. J. C. H. Watkins. *Learning with delayed rewards.* PhD thesis, University of Cambridge, Cambridge, UK, 1989.

Toward Seamless Transfer from Simulated to Real Worlds: A Dynamically–Rearranging Neural Network Approach

Peter Eggenberger[1], Akio Ishiguro[2], Seiji Tokura[2],
Toshiyuki Kondo[3], and Yoshiki Uchikawa[2]

[1] ATR Information Sciences Division,
Kyoto 619-0288, Japan
eggen@hip.atr.co.jp
[2] Dept. of Computational Science and Engineering, Nagoya University,
Nagoya 464-8603, Japan
ishiguro@cse.nagoya-u.ac.jp
tokura@cmplx.cse.nagoya-u.ac.jp
uchikawa@cse.nagoya-u.ac.jp
[3] Dept. of Computational Intelligence and Systems Science,
Tokyo Institute of Technology
Yokohama 226-8502, Japan
kon@dis.titech.ac.jp

Abstract. In the field of evolutionary robotics artificial neural networks are often used to construct controllers for autonomous agents, because they have useful properties such as the ability to generalize or to be noise–tolerant. Since the process to evolve such controllers in the real–world is very time–consuming, one usually uses simulators to speed up the evolutionary process. By doing so a new problem arises: The controllers evolved in the simulator show not the same fitness as those in the real–world. A gap between the simulated and real environments exists. In order to alleviate this problem we introduce the concept of neuromodulators, which allows to evolve neural networks which can adjust not only the synaptic weights, but also the structure of the neural network by blocking and/or activating synapses or neurons. We apply this concept to a peg–pushing problem for $Khepera^{TM}$ and compare our method to a conventional one, which evolves directly the synaptic weights. Simulation and real experimental results show that the proposed approach is highly promising.

1 Introduction

The *Evolutionary Robotics*(ER) approach has been attracting a lot of concern in the field of robotics and artificial life. In contrast to conventional approaches where designers have to construct robot controllers in a top–down manner, the methods in the ER approach can automatically construct controllers by taking *embodiment*(e.g. physical size of robot, sensor/motor properties and disposition, etc.) and the *interaction between the robot and its environment* into account.

J. Wyatt and J. Demiris (Eds.): EWLR 1999, LNAI 1812, pp. 44–60, 2000.
© Springer-Verlag Berlin Heidelberg 2000

In the ER approaches, artificial neural networks are widely used to construct controllers for autonomous mobile agents, because they can generalize, are non–linear and noise–tolerant [Floreano94,Nolfi97,Beer89,Ackley92]. Another advantage of neural network–driven robots is that a neural network is *low–level description* of a controller. More precisely, it directly maps sensor readings onto motor outputs. Although the ER approach has the above advantages, the following drawback still exist:

As the evolution in the real world is time–consuming, simulations are used to evolve the controller in a simulated environment and the best individuals are tested in the real world. The flaw of this combined approach is that evolved agents in simulated environments show often a significantly different behavior in the real world due to unforeseen perturbations. In other words, a *gap* between the simulated and real environments exists.

Therefore, evolved controllers should adapt not only to specific environments, but should be robust against environmental perturbations. Conventionally, to realize this requirement, many authors have been using techniques such as fitness–averaging [Reynolds94] and/or adding noise [Jacobi95,Miglino95] in the evolutionary process.

However, we speculate that these *passive* approaches are not essential solutions for the above requirements. We envision that without establishing a graceful method to realize robust controllers, we can not seamlessly transfer evolved agents from a simulator to the real world.

This leaves the following question. How can robots recognize their current situation and regulate their behavior appropriately?

A part of the answer may be that in the studies so fare made in ER no attempt was made to select directly for adaptation by changing the settings of the experiments. As even a simple thermostat needs sensory feedback to be able to control the temperature, an essential ingredient to any adaptive controller are controlling sensors to give the neural controller data how to change its current state towards the "good" one.

To construct robust controllers against environmental changes, in this study we focus on creation of feedback loops and their regulation mechanisms as the target to be evolved instead of evolving the synaptic weights (see Figure 1). If we can successfully evolve the appropriate regulation mechanism, we can expect high robustness against environmental perturbations.

In principle the information carried by the feedback loops can have the following two effects: Either one changes the weights of the synapses and the neurons' thresholds or one alters dynamically the structure of the neural network itself. The question is how can this be done and can such methods be used to solve the above problem.

Interestingly, neuroscientific results suggest that biological networks not only adjust the synaptic weights, but also the neural structure by blocking or activating synapses or neurons by the use of signaling molecules, so called *neuromodulators* [Meyrand91]. These findings stem from investigations made with the lobster's stomatogastric nervous system in which certain active neurons diffuse neuromodulators which then rearrange the networks. Note that the effect of a neuromodulator depends not only on theses substances, but also on the specific receptors, which are differently expressed in different cells. Imagine two cells A and B which expresses two different receptors C and D. A neuromodulator N will then only influence the cell(or synapse) which expresses the receptor A and not the cell with receptor B. The effect on a cell depends therefore on the interaction between the neuromodulator and the receptor and not just the neuromodulator alone.

The release of the neuromodulators depends on the activity of the neurons and therefore different sensor inputs may cause different patterns of released neuromodulators. As such dynamic mechanisms yield remarkable adaptation in living organisms, the proposed approach not only carries promise for a better understanding of adaptive networks, but they can be also applied to real–world problems as we already showed in previous work [Ishiguro99,Kondo99].

In this study, we apply the proposed concept to a peg–pushing problem for $Khepera^{TM}$ and compare our approach to a conventional one, which evolves directly the synaptic weights. As there exists no theory about how such dynamic neural network can be constructed, the evolutionary approach is the method of choice to explore the interactions between the neuromodulators, receptors, synapses and neurons.

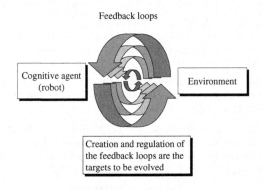

Fig. 1. Feedback loops between the robot and its environment.

2   Related Work

Although other authors tried to overcome the gap between simulations and real–world environments by fitness–averaging [Reynolds94] and/or adding noise [Jacobi95,Miglino95], these methods are quite passive and do not allow to tune the behavior specifically using differences in the sensory input.

Floreano et al. evolved neural controllers by assigning learning rule to every synapse [Floreano96]. In contrast, our approach allows to evolve learning rules, which can correlate not only the activity of neighboring neurons and their synapses, but can correlate neuronal activity from distant cells with synapses depending on which neuromodulators are diffused in a given situation. Furthermore, by introducing a blocking function, the neuromodulators can also dynamically rearrange the structure of the neural network.

Husbands et al. introduced a gas model (nitric oxide), which allows to modulate synapses [Husbands98]. The main difference to our approach is that we use specific receptors to locate an effect on a neuron or a synapse. Using the receptor concept a cell can diffuse the neuromodulator to all cells, but only those, which have the corresponding receptor will be changed by the neuromodulator, all the others rest unchanged.

3 Lessons to Be Learned from the Lobsters' Stomatogastric Nervous System

3.1 Dynamic Rearrangement

Investigations carried out on the lobsters' stomatogastric nervous system suggest that biological nervous systems are able to dynamically change their structure as well as their synaptic weights [Meyrand91].

This stomatogastric nervous system mainly consists of an *oesophageal*, a *pyloric*, and a *gastric* network. Normally, these three individual networks show their own independent oscillatory behaviors, but in the moment a lobster is eating the networks are integrated and reconstructed to a new one, the swallowing network, in which certain neurons and connections are excluded and formerly inactive connections are activated (see Figure 2).

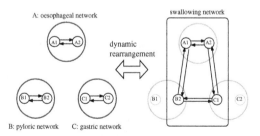

Fig. 2. Dynamic rearrangement of a lobster's stomatogastric nervous system.

Recent studies in neurophysiology showed neuromodulators(hereafter: NMs) play a crucial role to regulate this remarkable phenomenon(e.g. changing properties of synapses as well as neurons).

3.2   Neuromodulators

NMs are substances that can dynamically influence several properties of neurons and therefore the function of a neural network. In contrast to neurotransmitters(NTs) the effect of NMs spreads slower and lasts longer. NMs change the processing characteristics of neural networks by affecting the membrane potential, the rate of changing the synapses(i.e. influence on learning mechanisms) and other parameters. Typical NMs are *acetylcholine, norepinephrine, serotonin, dopamin*(all are also used as NTs), *somatostatine* and *cholecystokinine*(both also used as hormones in the human body) and many small proteins. Although these substances are released in a less local manner than NTs, the effects can be quite specific. This specificity comes from specific receptors on the neurons and their synapses.

These NMs stem either locally from the neural network itself or from specific sub–cortical nuclei. The local release of NMs depends on the activity of the local neural network itself. On the other hand, sub-cortical nuclei as the locus coeruleus (*noradrenergic innervation*), the ventral tegmental area(*dopaminergic innervation*) or the basal fore–brain nuclei(*cholinergic innervation*) send neuromodulatory axons to cortical structures to release NMs from axonal varicosities which is called volume transmission. Many publications in neuroscience show the importance of NMs for dynamic rearrangement of neuronal modules [Meyrand91,Hooper89] or for learning and memory(switching between learning and recall mode) [Hasselmo95].

In this study we implemented the following properties:

- dynamic change of the threshold of a neuron
- dynamic blocking of neurons and synapses
- dynamic change of the inhibitory or excitatory properties of a synapse
- dynamic modulation of the synaptic weights (i.e. learning).

4   Proposed Method

4.1   Basic Concept

The basic concept of our proposed method is schematically depicted in Figure 3. As in the figure, unlike the conventional neural networks, we assume that each neuron can potentially diffuse its specific (i.e. genetically–determined) NMs according to its activity, and each synapse has receptors for the diffused NMs. We also assume that each synapse independently interprets the received NMs, and changes its properties(e.g. synaptic weight). By selecting for regulatory feedback loops(cyclical interaction between the diffusion and reaction of NMs), we expect to be able to evolve adaptive neural networks, which show a seamless transfer from simulations to the real world(in the figure, the thick and thin lines denote the connections being strengthened and weakened by NMs, respectively).

In summary, in contrast to the conventional ER approach that evolves synaptic weights and neuron's bias of neuro–controllers, in this approach we evolve the following mechanisms:

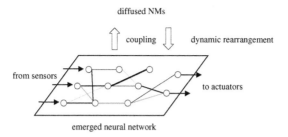

Fig. 3. Basic concept.

- Diffusion of NMs(when, which type of NMs are diffused from each neuron?)
- Reaction to NMs(how do the receptors on each synapse interpret the received NMs?)

To determine the above parameters, we use a Genetic Algorithm(GA).

4.2 Application Problem

In this study, we use a *peg–pushing problem* as a practical example. Figure 4 schematically shows this problem. As in the figure, in this environment there are three objects: the light source, peg and robot. Here the task of the robot is to push the peg toward the light source. Due to the rudimentary stage of investigation, the following are not taken into account:

- Collision to the surrounded walls (in such a case, the trial is terminated).
- The robot contacts with the peg at the initial position.

Figure 5 illustrates the structure of the robot used in this experiment. This robot is a simulated $Khepera^{TM}$ robot, equipped with six infra-red sensors in

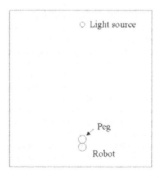

Fig. 4. A peg–pushing problem.

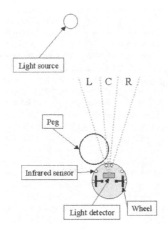

Fig. 5. Structure of the robot.

the frontal side, and light direction detector, and two DC motors. In this study, the following equation is used to simulate the property of the infra–red sensor [Hoshino94]:

$$Sensor\_value = \frac{1023}{1 + \exp(0.65d - 14cos(1.3\alpha))},\qquad(1)$$

where d represents the distance between the sensor and peg, and α denotes the angle between the direction of the sensor heading and the detected object.

On the other hand, the light direction detector can perceive the direction of the light source in three ranges as (see Figure 5):

$$Light\_Left = \begin{cases} 1.0\cdots -18° \le \theta \le 0° \\ 0.0\cdots otherwise \end{cases}$$

$$Light\_Center = \begin{cases} 1.0\cdots -9° \le \theta \le 9° \\ 0.0\cdots otherwise \end{cases}$$

$$Light\_Right = \begin{cases} 1.0\cdots 0° \le \theta \le 18° \\ 0.0\cdots otherwise. \end{cases}$$

$$(2)$$

4.3   Controller

Figure 6 shows the neural network structure for the robot controller. As in the figure, this controller consists of nine sensory neurons(i.e. six infra–red, and three light direction sensors), five hidden neurons and four motor neurons(i.e. CW/CCW motor neurons for the right and left motors).

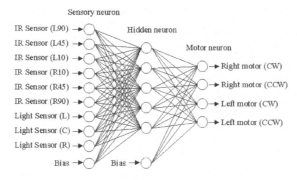

Fig. 6. Structure of the controller.

4.4 Diffusion of NMs

Figure 7 schematically illustrates the diffusion process. As in the figure, each neuron can diffuse its specific(i.e. genetically–determined) type of NMs in the case where its activation value is within the specific threshold values(these are also genetically–determined).

Here, for simplicity, we assume that each neuron can diffuse at most one type of NMs. Furthermore, we set the number of NM types to two.

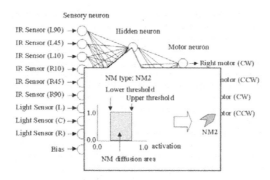

Fig. 7. Diffusion of the NMs.

4.5 Reaction to the Diffused NMs

As in Figure 8, each synapse and neuron has receptors for the diffused NMs. Once the NMs are diffused from certain neurons, these NMs may change the

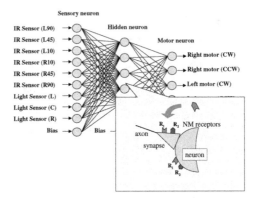

Fig. 8. Reaction to the diffused NMs.

property of each synapse and neuron. This change is carried out according to the genetically–determined interpretation. Detailed explanation is given below.

Dynamic change of synaptic property. Table 1 represents an example of the receptor interpretation for the change of synaptic property. Due to the number of NM types(i.e. two), all possible combination of the received NMs is four, in this case. Each case can take one of the following four types of modulation; *Hebbian learning* ($R_{ij}(NM)$=1.0), *anti–Hebbian learning* ($R_{ij}(NM)$=-1.0), *non–learning* ($R_{ij}(NM)$=0.0), and *blocking*(i.e. excluding the synapse), respectively.

Here, we use the following equation for the dynamic modulation of the synaptic weights.

$$w_{ij}^{t+1} = w_{ij}^{t} + \eta \cdot R_{ij}(NM) \cdot a_i \cdot a_j, \tag{3}$$

where, η is the leaning rate, and $R_{ij}(NM)$ is the parameter which determines how the synapse concerned interprets its received NMs and modifies its weight (see above).

Table 1. An example of the NM–interpretation table for the change of synaptic property.

NM_1	NM_2	Type of modulation
0*	0	Hebbian learning
0	1	non–learning
1**	0	blocking
1	1	anti–Hebbian learning

* 0 means NM_n is not diffused.

** 1 means NM_n is diffused.

Table 2. An example of the NM–interpretation table for the change of neuronal property.

NM_1	NM_2	Type of modulation
0*	0	active
0	1	inactive
1**	0	inactive
1	1	active

* 0 means NM_n is not diffused.
** 1 means NM_n is diffused.

Dynamic change of neuronal property. Table 2 depicts an example of the receptor interpretation for the change of neuronal property. Each case can take one of two state; active and inactive. This assignment is determined through the evolutionary process.

In the following, we show simulated and experimental results, and compare the proposed method to a conventional one, which directly evolves the synaptic weights. Due to the rudimentary stage of investigation, blocking/activating synapses and neurons, and receptors on neurons are not implemented here. Incorporating these mechanisms is currently proceeding.

5 Simulation Results

5.1 Encoding

As illustrated in figure 9, the genetic information has two main parts: a diffusion part and a reaction part(the genetic information is encdoded in a binary string). The diffusion part encodes the types of neuromodulators a neuron can release and when(determined by two thresholds) a neuromodulator will be released. The diffusion part contains 18 blocks of information for the 18 neurons used in our neural network. The thresholds are encoded as 8 bits whereas for the NM type 2 bits are used.

The reaction part determines the receptor types(up to two) of a neuron or synapse and which function(block, learning and non–learning) will be applied in the case one or two receptors become active. As we use for the moment just two receptors in our model, we need two bits to encode the information contained in Table 1.

5.2 Evaluation Criterion and Conditions

In order to verify the validity of our proposed method, in the following we compare it with the conventional ER approach where the synaptic weights are the target to be evolved. The evaluation criterion used for the evolutionary process is:

$$fitness = \left\{ 1 - \frac{dist(light, peg_{end})}{dist(light, peg_{start})} \right\}^2 \times 100, \tag{4}$$

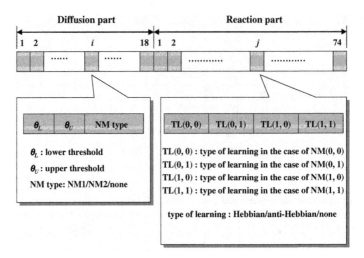

Fig. 9. Encoding scheme for the dynamically-rearranging neural network.

Table 3. Parameters used in the simulation.

Learning rate (η)	0.1
Population	100
Generation	200

where, $dist(light, peg_{start})$ is the distance between the light source and the initial peg position. Similarly, $dist(light, peg_{end})$ represents the resultant distance between the light source and the resultant peg position.

Each individual is allowed to move during 500 time–steps in the environment shown in Figure 4. Furthermore, each individual is tested ten times from the different random heading direction. Thus, the averaged fitness value of each individual is used for the evaluation. The parameters used in the following simulations are listed in Table 3.

During the evolutionary process, tournament selection scheme, uniform crossover and random mutation are adopted.

5.3   Results

Figure 10 shows the resultant fitness transition of both methods. As in the figure, both methods finally obtain high scores as the evolutionary process iterated. Figure 11 depicts typical trajectories of the best agents obtained in the final generation. From the figures, it is understood that both agents can successfully carry the peg toward the goal.

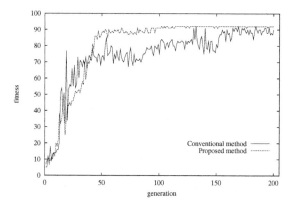

Fig. 10. Comparison of the fitness transition.

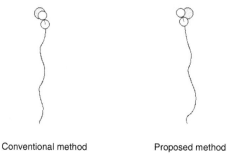

Conventional method   Proposed method

Fig. 11. Resultant trajectories of the best agents in both methods.

5.4 Evaluation of Adaptability

As mentioned earlier, the evolved agent should be not only adapted to the given environments during the evolutionary process, but also be robust against environmental perturbations. Without this we can not seamlessly tranfer the evolved agents from the simulated to the real worlds. In the following, we show the quantitative comparison of adaptability of both methods.

Robustness against the specific environmental perturbations. In order to verify the robustness, we observed the behavior of the evolved agents in the following test environments:

Table 4 shows the results of these tests. Each result is represented by the fitness value averaged over one hundred trials. From these results, it is understood that the proposed method yields remarkable high scores while the conventional method shows serious degradations.

Test1: Add 10 % slip to the right motor output.
Test2: Add 10 % slip to the left motor output.
Test3: Add 0.2 degree to the direction of
        the resultant peg movement.
Test4: Add -0.2 degree to the direction of
        the resultant peg movement.

Table 4. Quantitative comparison of the robustness against the specific environmental perturbation.

	Conventional method	Proposed method
Without Perturbation	83.0	96.1
Test 1	66.6	92.6
Test 2	70.3	89.8
Test 3	71.5	79.6
Test 4	77.7	80.1

Robustness against the alteration of the peg size. Figure 12 shows the result of the comparison of the robustness against the alteration of the peg size. In the figure, horizontal axis represents the magnification of the peg size(e.g. 0.5 means a half size of the original peg). From this result, we can see that the proposed method can keep high fitness value over these alterations compared with the conventional method.

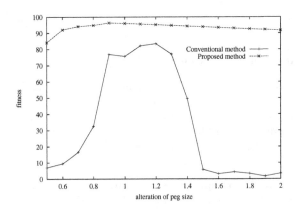

Fig. 12. Quantitative comparison of the robustness against the alteration of the peg size.

Table 5. Quantitative comparison of the robustness against the random noise.

	Conventional method	Proposed method
Without Noise	83.0	96.1
With Noise	75.1	92.2

Robustness against random noise. We furthermore observed the resultant behaviors of both methods by adding 5% and 3% random noises based on the Gaussian distribution onto each motor output and direction of pushed peg movement, respectively.

Table 5 shows the results of this test averaged over one hundred trials. From the table the proposed shows high adaptability. Figure 13 is a typical example of the resultant trajectories of both methods. The robot driven by the dynamically–rearranging neural network can successfully reach the goal. Figure 14 is an ex-

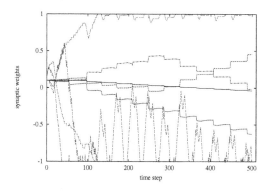

Conventional method   Proposed method

Fig. 13. Typical resultant trajectories.

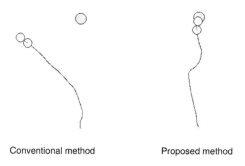

Fig. 14. An example of the transition of the synaptic weights.

ample of the transition of the synaptic weights. In the figure, for convenience several synapses are selected. Interestingly, some synapses drastically change their weights from excitatory to inhibitory and *vice versa*, and others settle to certain values. From this result, we speculate that the neural networks suitable for the current situation are emerged by appropriately changing the correlation among neurons. Detailed analysis is currently under investigation.

6   Experimental Results

As mentioned in the previous section, the proposed dynamically–rearranging neural network shows high adaptability against the environmental perturbations compared with the conventional ER approach where the synaptic weights are the target to be evolved. This implies the proposed method would be highly promising to reduce the serious gap between simulated and real environments.

In the following, we investigate how the evolved controllers in both methods work after tranferring onto the real mobile robot. Figure 15 and 16 show the real robot and its experimental setup, respectively. Figure 17 depicts the experimental results of the evolved agents which are identical to the ones in Figure 10. Unlike in the simulated environments, the agent based on the conventional method cannot successfully carry the peg. On the other hand, the agent of the proposed method show the same function as in the simulation.

7   Conclusion

In this paper, we proposed a new adaptive controller for autonomous mobile robots by incorporating the concept of dynamic rearrangement of neural networks with the use of neuromodulators.

We investigated the adaptability of the proposed method in comparison with the conventional ER approach by carrying out simulations. The obtained results are encouraging. Detailed analysis of the obtained neural networks(e.g. how do the NMs change the correlation among neurons?) are currently under investigation.

Fig. 15. Experimental robot ($Khepera^{TM}$).

Fig. 16. Experimental setup.

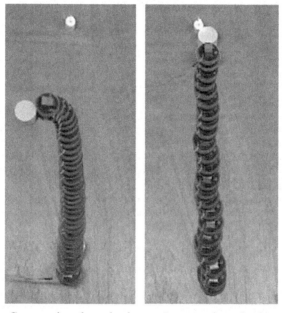

Conventional method   Proposed method

Fig. 17. Experimental results.

We are also concurrently investigating the implementation of the proposed dynamically–rearranging neural networks by FPGAs(Field Programmable Gate Arrays). Full implementation of the evolved dynamically–rearranging neural networks is currently proceeding.

References

Floreano94. Floreano, D. and Mondada, F. (1994). Automatic creation of an autonomous agent: Genetic evolution of a neural-network driven robot, *Proc. of the 3rd International Conference on Simulation of Adaptive Behavior*, MIT Press, pp.421–430

Floreano96. Floreano, D. and Mondada, F. (1996). Evolution of Plastic Neurocontrollers for Situated Agents, *From animals to animats 4: Proc. of the 4rd International Conference on Simulation of Adaptive Behavior*, MIT Press, pp.401-411

Nolfi97. Nolfi, S. Parisi, D. (1997). Learning to adapt to changing environments in evolving neural networks, *Adaptive Behavior*, 5-1, pp.75–98

Reynolds94. Reynolds, C. W. (1994). An Evolved, Vision-Based Model of Obstacle Avoidance Behavior, *ARTIFICIAL LIFE III*, pp.327–346

Jacobi95. Jacobi, N. Husbands, P. and Hervey, I. (1995). Noise and the Reality Gap: The Use of Simulation in Evolutionary Robotics, *Third European Conf. on Artificial Life (ECAL95), Advances in Artificial Life*, pp.704–720, Springer

Husbands98. Husbands, P. Smith, T. O'Shea, M. Jakobi, N. Anderson, J. and Philippides, A. (1998). Brains, Gases and Robots. *In Proc. ICANN98*, pp 51-63, Springer–Verlag.

Miglino95. Miglino, O. Lund, H. H. and Nolfi, S. (1995). Evolving Mobile Robots in Simulated and Real Environments, *Artificial Life 2*, pp.417–434

Beer89. Beer, R. Chiel, J. and Sterling, L. (1989). An artificial insect, *American Scientist*, **79**, pp.444–452

Ackley92. Ackley, D. Littman, M. (1992). Interactions Between Learning and Evolution, *Artificial Life II*, Addison-Wesley, pp.487–509

Meyrand91. Meyrand, P. Simmers J. and Moulins, M. (1991). Construction of a pattern-generating circuit with neurons of different networks, *NATURE*, **351**-2MAY, pp.60–63

Hooper89. Hooper, S. L. and Moulins, M. (1989). Switching of a Neuron from One Network to Another by Sensory-Induced Changes in Membrane Properties, *SCIENCE*, **244**, pp.1587–1589

Hasselmo95. Hasselmo, M. (1995). Neuromodulation and cortical function: modeling the physiological basis of behavior, *Behavioral Brain Research*, **67**, Elsevier Science B.V., pp.1–27

Hoshino94. Hoshino, T (1994). Dreams and Worries of Artificial Life, *Popular Science Series*, Shouka-Bo (in Japanese)

Ishiguro99. Ishiguro, A. Kondo, T. Uchikawa, Y. and Eggenberger, P. (1999). Autonomous Robot Control by a Neural Network with Dynamically-Rearranging Function, *Proc. of the 11th SICE Symposium on Decentralized Autonomous Systems*, pp.213–218 (in Japanese)

Kondo99. Kondo, T. Ishiguro, A. Uchikawa, Y. and Eggenberger, P. (1999). Autonomous Robot Control by a Neural Network with Dynamically-Rearranging Function, *Proc. of the 4th International Conference on Artificial Life and Robotics (AROB99)*, 1, pp.324–329

How Does a Robot Find Redundancy by Itself?
- A Control Architecture for Adaptive Multi-DOF Robots -

Koh Hosoda* and Minoru Asada

Department of Adaptive Machine Systems, Osaka University, Yamadaoka 2–1, Suita,
Osaka 565–0871, Japan
hosoda@ams.eng.osaka-u.ac.jp

Abstract. Finding redundancy with respect to a given task in an environment is important ability for a robot to dynamically assign the degrees of freedom to several tasks. In this paper, an adaptive controller is proposed that does not only estimate appropriate control parameters but also can find redundancy with respect to the given task dynamically. The controller is derived based on a least-mean-square method. A simulation result of a camera-manipulator system is shown to demonstrate that the proposed method can find redundancy automatically.

1 Introduction

General robots nowadays, such as industrial robots, have minimum number of degrees of freedom and minimum variety of sensors only sufficient for presupposed situations. Such robots can achieve restricted tasks properly in restricted environments. However, they cannot deal with unexpected situations appropriately since they are equipped with only necessary and sufficient numbers of actuators and sensors.

A robot, as a universal machine, ought to have adaptivity, ability to estimate appropriate control parameters and/or structure to achieve a given task in an environment. So as to have such adaptivity against changes of task and environment, a robot needs to have larger number of actuators and more variety of sensors. Such redundancy even can allow us to design a robot so that new behaviors can emerge [1]. In this sense, we have to focus on a multi-DOF robot that has many actuators and a variety of sensors.

In figure 1(a), a general control architecture for a less-DOF robot is shown. Sensor data is compared with a given task and is fed into the controller. Then, the controller calculates commands for actuators. Just by extending this way, we can derive a controller for a multi-DOF robot(figure 1(b)) although this can only achieve a single task. The robot may have more ability if it can achieve several tasks at a time. In addition, the robot will be more adaptive if it can *dynamically* assign its degrees of freedom to each task since this means the robot can estimate not only appropriate parameters but also an appropriate control structure(figure 1(c)). To assign degrees of freedom dynamically, it is important

---

* This work is partly done during his stay in the University of Zürich.

J. Wyatt and J. Demiris (Eds.): EWLR 1999, LNAI 1812, pp. 61–70, 2000.

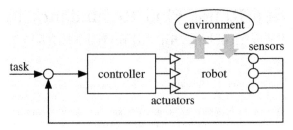

(a) A control architecture for a robot with several degrees of freedom

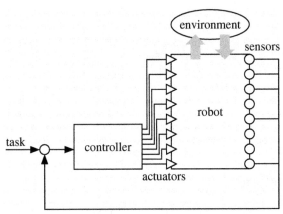

(b) A control architecture for a robot with more degrees of freedom although it cannot find redundancy automatically

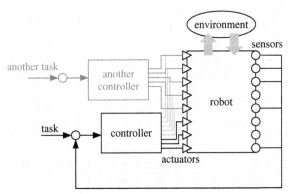

(c) A control architecture for a robot with more degrees of freedom with ability to find redundancy and to spare it for another task

Fig. 1. A control architecture for a less-DOF robot (a) is extended for a multi-DOF robot although it can achieve only one task (b). If the controller can assign DOF to the given task dynamically, it can spare the remainder for another task.

for the controller to be able to find redundancy with respect to the task on-line since necessary number of degrees for a task depends on the environment where it is given. Even in case of malfunction of several actuators, the robot can assign unbroken actuators to the task, and the task will be achieved.

There has been a certain amount of work trying to make an adaptive robot. In the field of control theory, they have developed adaptive control schemes which can estimate parameters but cannot find redundancy [2], [3]. There are also several attempts to make an adaptive robot by a control architecture such as a subsumption architecture [4] and a schema system [5], each of which also cannot utilize redundancy of the robot effectively.

Our final goal is to build an autonomous adaptive robot that has many degrees of freedom and sensors and that can utilize redundancy to adapt to a dynamic environment. We have proposed to use a hybrid structure of adaptive controllers for multi-DOF robots [6], [7], [8]. In these papers, each controller can estimate parameters, but cannot find redundancy with respect to a given task, and therefore cannot utilize it.

In this paper, an adaptive controller is proposed which estimates not only parameters but also sufficient degrees of freedom for a task. The remainder of this article is organized as follows. First, the adaptive controller is introduced based on a least-mean-square method. Then, by utilizing UD factorization of the covariance matrix, a method to derive the redundancy is proposed. Finally, a simulation result is shown to demonstrate the effectiveness of the proposed method.

2   Adaptive Controller That Can Find Redundancy

2.1   Task Definition and Feedback Controller

The i-th task for the robot is assumed to be defined in the i-th sensor space: $x_i \rightarrow x_{id}$. The class of tasks is limited by this definition, but still a certain amount of tasks can be defined in this way: for example, to track a moving visual target [9], to keep balance of a legged robot [7], to exert force towards constraint surface [6], to maximize manipulability [10], and so on.

The sensor output x_i is a function of actuator displacement θ,

$$x_i = x_i(\theta). \tag{1}$$

Differentiating eq.(1) with respect to the sampling rate, we can get

$$\Delta x_i(k) = J_i(k)\Delta\theta(k), \tag{2}$$

in the k-th sampling step, where J_i is a Jacobian matrix $(= \partial x_i/\partial \theta^T)$. Assume that all the actuators of the robot are velocity-controlled, and we can derive a velocity command for the i-th task from eq.(2), based on well-known resolved rate control [10]

$$u(k) = J_i(k)^+ K_i(x_{id}(k) - x_i(k)) + (I - J_i(k)^+ J_i(k))\xi, \tag{3}$$

where $\mathbf{I}$, $\mathbf{K}_i$ and $\boldsymbol{\xi}$ are an identity matrix, a feedback gain matrix, and an arbitrary vector that denotes the redundancy of the controller with respect to the given task to follow $\boldsymbol{x}_{id}$. $\boldsymbol{A}^+$ denotes a pseudo-inverse of a matrix $\boldsymbol{A}$. By utilizing the last term in the right hand side of eq.(3), we can make a hybrid structure of controllers [6]. If the robot and the environment are well-calibrated and do not change dynamically, the Jacobian matrix $\boldsymbol{J}_i(k)$ is known, and therefore, the resolved rate controller (3) can achieve the task. In case they are not calibrated, or the environment changes dynamically, the matrix $\boldsymbol{J}_i(k)$ is unknown, and we have to estimate it.

2.2   On-Line Estimation of Jacobian Matrix

To estimate the matrix $\boldsymbol{J}_i(k)$ the authors have proposed a method based on the least-mean-square method [6]:

$$\widehat{\boldsymbol{J}}_i(k) = \widehat{\boldsymbol{J}}_i(k-1) + \{\Delta\boldsymbol{x}_i(k) - \widehat{\boldsymbol{J}}_i(k-1)\Delta\boldsymbol{\theta}(k)\} \frac{\Delta\boldsymbol{\theta}(k)^T \boldsymbol{P}(k-1)}{\rho + \Delta\boldsymbol{\theta}(k)^T \boldsymbol{P}(k-1)\Delta\boldsymbol{\theta}(k)}, \quad (4)$$

where $(0 < \rho < 1)$ is a forgetting factor, and $\boldsymbol{P}(k)$ is a covariance matrix calculated as

$$\boldsymbol{P}(k) = \frac{1}{\rho}\left\{\boldsymbol{P}(k-1) - \frac{\boldsymbol{P}(k-1)\Delta\boldsymbol{\theta}(k)\boldsymbol{\theta}(k)^T\boldsymbol{P}(k-1)}{\rho + \Delta\boldsymbol{\theta}(k)^T\boldsymbol{P}(k-1)\Delta\boldsymbol{\theta}(k)}\right\}. \quad (5)$$

We have already demonstrated that many kinds of robots can be controlled utilizing the estimator (5) and the feedback controller (3) without any *a priori* knowledge on the robot nor on the environment [9].

2.3   Estimating Redundancy

Utilizing a characteristic of the estimator (5), it can find the redundancy with respect to the task. If the robot is redundant to achieve the i-th task, the variation of the actuator displacement $\Delta\boldsymbol{\theta}$ is not uniform over the task space. Therefore, by observing the covariance matrix $\boldsymbol{P}$ which represents the variation of $\Delta\boldsymbol{\theta}$ the controller can find redundancy.

It is known if the variation of $\Delta\boldsymbol{\theta}$ is not uniform, the covariance matrix $\boldsymbol{P}$ is no more positive, and therefore the least-mean-square based estimator will lose numerical stability [11]. To avoid such instability, a technique so called UD factorization is proposed [12], which we see more details in the appendix. Following the method, the covariance matrix $\boldsymbol{P}$ is factorized

$$\boldsymbol{P} = \boldsymbol{U}\boldsymbol{D}\boldsymbol{U}^T, \quad (6)$$

where $\boldsymbol{U}$ is an upper triangular matrix whose diagonal elements are all 1, and $\boldsymbol{D}$ is a diagonal matrix. Instead of updating $\boldsymbol{P}$ by eq.(5), we update $\boldsymbol{U}$ and $\boldsymbol{D}$ (see appendix) to ensure the positiveness of the matrix $\boldsymbol{P}$ and therefore, we can ensure the numerical stability.

Note that if the positiveness of P is weak, one of the diagonal elements of D is nearly 0. Therefore, if the control input for the robot $\Delta\theta$ is not uniform, we can find it by observing D. Let D and U be

$$D = \operatorname{diag}\left[d_1 \cdots d_n \right], \tag{7}$$

$$U = \begin{bmatrix} 1 & & * \\ & \ddots & \\ 0 & & 1 \end{bmatrix}$$

$$= \left[u_1 \cdots u_n \right], \tag{8}$$

respectively. If the robot is redundant with respect to the task, a diagonal element d_j of D becomes very small. If $|d_j| < \epsilon$ (ϵ is a considerably small number), the corresponding direction of $\Delta\theta$ is not anymore effective to the estimation, that is, the direction is redundant with respect to the task. We can easily find that in eq.(4) every term concerning about $\Delta\theta$ is multiplied with P. Since the term $P\Delta\theta$ is factorized,

$$P\Delta\theta = UDU^T\Delta\theta$$

$$= U \begin{bmatrix} \ddots & & \\ & d_j & \\ & & \ddots \end{bmatrix} U^T\Delta\theta,$$

the redundant direction is given as a column vector of U^{-T}. To eliminate the redundancy, we remove this direction from the estimated matrix $\widehat{J}_i$.

3 Simulation

We show a simulation result to demonstrate that the proposed scheme can find redundancy with respect to one task, and that the robot can apply another task by utilizing the redundancy.

A robot used for the simulation has 4 DOFs with two cameras depicted in figure 2. By two cameras the robot observes two positions of image points in the image planes. The first task given to the robot is to keep these positions constant in the image planes, that is, a visual servoing task. The points are moving along circles on a plane whose diameter is 0.1[m] in 5[s], 10 times.

Two cameras are gazing at 2 points, therefore the image feature vector is $x \in \Re^8$. The robot has 4 DOFs, $\theta \in \Re^4$. Two points are moving along the circle on a plane, therefore the DOF needed for the task is 3.

The forgetting factor of the estimator $\rho = 0.8$. The initial matrix of P is an identity matrix. The initial matrix for the estimated Jacobian matrix is given as

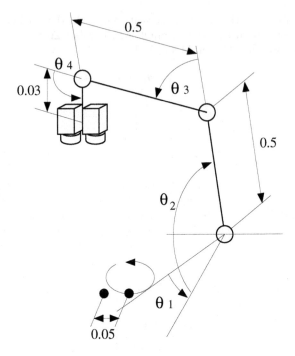

Fig. 2. A 4 DOF manipulator with 2 cameras is used for the simulation (Unit is [m]). Each of visual targets(two points) is moving in a circle on a plane.

an arbitrary matrix

$$\hat{J}(0) = \begin{bmatrix} 100 & 0 & 0 & 0 \\ 0 & 100 & 0 & 0 \\ 0 & 0 & 100 & 0 \\ 0 & 0 & 0 & 100 \\ 100 & 0 & 0 & 0 \\ 0 & 100 & 0 & 0 \\ 0 & 0 & 100 & 0 \\ 0 & 0 & 0 & 100 \end{bmatrix},$$

because we assume that the controller does not have any *a priori* knowledge on the parameters of the robot and of the environment. Note that the rank of this matrix is 4. This means that all the 4 DOFs are utilized for the visual servoing task at the beginning of the control period.

To help understanding of the readers, we show the approximate matrix of true Jacobian:

$$\widehat{\boldsymbol{J}}(0) = \begin{bmatrix} 2300 & 0 & 0 & 0 \\ 100 & 0 & 1300 & 1300 \\ 2400 & 300 & 200 & 0 \\ 0 & 0 & 1300 & 1300 \\ 2400 & -300 & -200 & 0 \\ 100 & 0 & 1300 & 1300 \\ 2300 & 0 & 0 & 0 \\ 0 & 0 & 1300 & 1300 \end{bmatrix}.$$

Note that the rank of this matrix is 3.

Here we show two simulation cases:

case 1 applying visual servoing control without the redundancy estimator, and
case 2 applying visual servoing control with the redundancy estimator.

In case 1, the estimator cannot find redundancy with respect to the visual servoing task whereas it can in case 2. In case 2, utilizing the redundancy, the second task to make θ_3 converge to $-\pi/3$ is applied. It is performed by minimizing the index

$$Q = \{\theta_3 - (-\pi/3)\}^2.$$

A simulation result is shown in figures 3 and 4. From figure 3, we can say that the performances in the image planes in both cases are almost the same (hardly distinguishable). In both cases, the visual servoing task is achieved without knowing about structure and parameters of the robot and the environment by utilizing the on-line estimator. In figure 4, around 3 [s], the proposed estimator finds the redundancy with respect to the visual servoing task. After that, θ_3 is converging to $-\pi/3$ utilizing the redundancy(see the dashed line) whereas the controller cannot utilize redundancy in case 1.

4   Conclusions and Discussion

In this paper, an adaptive control scheme has been proposed that can estimate not only parameters but also sufficient degrees of freedom to achieve a given task. It can find redundancy with respect to the task by itself and can spare it for other tasks. By combining such controllers, it is easy to construct a hybrid control architecture for a robot that has many actuators and sensors. At first, only the first task controller is working, then it finds redundancy with respect to the first task. It will spare the found redundancy for the second task controller, and so forth. Note that the first controller does not simply inhibit the lower controllers, but spares the found redundancy for them. As a consequence, a behavior is emerged by a combination of several controllers without any supervisor.

For example, to control a humanoid that has many degrees of freedom and sensors, the effort to program all the degrees grows enormous. In addition, the environmental constraints are difficult to be predicted. For such a robot, the proposed method is supposed to be very powerful. One should prepare several tasks

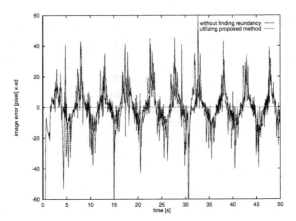

Fig. 3. Simulation result 1 : image error in camera 1, image feature 1, x- coordinate $(x_1 - x_{1d})$

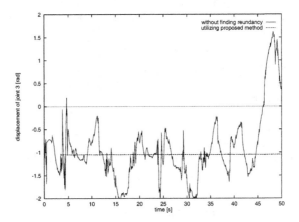

Fig. 4. Simulation result 2 : joint 3 displacement θ_3

to be achieved, and apply the proposed scheme to each task. Since each controller will automatically find redundancy in the relation with the environment and spare it for other task, explicit coordination is not needed.

In the proposed method to construct the hybrid control structure, it is necessary that each of tasks must be defined by the desired sensor output x_d (or $x_d(t)$, a function of time), which obviously limits the applicability of this method. However, a certain amount of tasks still can be defined in this way: for example, to track a moving visual target [9], to keep balance of a legged robot [7], to exert force towards a constraint surface [6], to maximize manipulability [10], and so on. This means there are a certain amount of behaviors that can be emerged by the proposed method.

Acknowledgements

One of authors, Koh Hosoda, would like to thank Prof. Rolf Pfeifer and his colleagues who made valuable discussion with him, since this work is partly done during his stay in AI–lab, the University of Zürich. He also would like to thank Murata overseas scholarship to support the stay.

References

1. R. Pfeifer and C. Scheier. *Understanding Intelligence*, chapter 10 : Design Principles for Autonomous Agents. The MIT Press, 1999.
2. Y. D. Landau. *Adaptive Control —The Model Reference Approach*. Marcel Dekker, 1979.
3. J. J. Slotine and W. Li. Adaptive manipulator control: A case study. *IEEE Transaction on Automatic Control*, 33(11):995–1003, 1988.
4. R. A. Brooks. Intelligence without representation. *Artificial Intelligence*, 47, 1991.
5. R. C. Arkin. *Behavior-Based Robotics*. The MIT Press, 1998.
6. K. Hosoda, K. Igarashi, and M. Asada. Adaptive hybrid control for visual servoing and force servoing in an unknown environment. *IEEE Robotics and Automation Magazine*, 5(4):39–43, 1998.
7. T. Miyashita, K. Hosoda, and M. Asada. Hybrid structure of reflective gait control and visual servoing for walking. In *Proc. of the 1998 IEEE/RSJ Int. Conf. on Intelligent Robots and Systems*, pages 229–234, 1998.
8. T. Miyashita, K. Hosoda, and M. Asada. Reflective walk based on lifted leg control and vision-cued swaying control. In *Proc. of 1998 BSMEE International Symposium on Climbing and Walking Robots (CLAWAR'98)*, pages 349–354, 1998.
9. K. Hosoda and M. Asada. Adaptive visual servoing for various kinds fo robot systems. In A. Casals and A. T. de Almeida, editors, *Experimental Robotics V*, pages 547–558. Springer, 1998.
10. T. Yoshikawa. *Foundations of Robotics*. The MIT Press, 1990.
11. E. C. Ifeachor and B. W. Jervis. *Digital Signal Processing – A Practical Approach –*. Addison-Wesley, 1993.
12. G. J. Bierman. Measurement updating using the U–D factorization. *Automatica*, 12:375–382, 1976.

A   UD Factorization [12]

If the initial UD factorization of P is given, updated matrices $\widehat{U}$ and $\widehat{D}$ are obtained from U and and D:

$$f = U^T \Delta\theta,$$
$$v_i = d_i f_i \qquad (i = 1, \cdots, n, n \text{ is dimension of } P),$$
$$\alpha_1 = \rho + v_1 f_1,$$
$$\hat{d}_1 = d_1 \gamma / \alpha_1,$$
$$b_1 = v_1,$$

As for the $j = 2, \cdots, n$ -th elements, they are updated as follows, where $\stackrel{\triangle}{=}$ denotes overwite in the program manner,

$$\alpha_j = d_{j-1} + f_j v_j,$$
$$\hat{d}_j = d_j \alpha_{j-1}/\alpha_j,$$
$$b_j \stackrel{\triangle}{=} v_j,$$
$$p_j = -f_j/\alpha_{j-1},$$

$$\left.\begin{array}{l} \widehat{U}_{ij} = U_{ij} + b_i p_j \\ b_i \stackrel{\triangle}{=} b_1 + U_{ij} v_j \end{array}\right\} (i = 1, \cdots, j). \tag{9}$$

$$d_i \stackrel{\triangle}{=} d_i/\rho \qquad (i = 1, \cdots, n). \tag{10}$$

Learning Robot Control by Relational Concept Induction with Iteratively Collected Examples

Nobuhiro Inuzuka, Taichi Onda*, and Hidenori Itoh

Department of Intelligence and Computer Science,
Nagoya Institute of Technology,
Gokiso-cho, Showa-ku, Nagoya 466-8555, Japan
{inuzuka,taichi,itoh}@ics.nitech.ac.jp

Abstract. This paper investigates a method for collecting examples iteratively in order to refine the results of induction within the framework of inductive logic programming (ILP). The method repeatedly applies induction from examples collected by using previously induced results. This method is effective in a situation where we can only give an inaccurate teacher. We examined this method by applying it to robot learning, which resulted in increasing the refinement of induction. Our method resulted in the action rules of a robot being learned from a very rough classification of examples.

Keywords: robot learning, ILP, collection method of examples, inaccurate teachers.

1   Introduction

Learning control rules of mobile robots is an important application in the field of machine learning (ML), in which many methods are being investigated. A direct application of concept learning to induce control rules of real mobile robots is difficult for reasons that include noisy sensor information of robots, and real time processing. Another interesting and important difficulty of robot learning is that we cannot have an accurate teacher for learning. We can have a model of robot behavior for short periods or step-wise behavior, but it is difficult to say how behavior affects a robot's situation after a while. A teacher must indicate if a robot's action is appropriate for a sequence of actions and even for a total goal. We cannot obtain this kind of teacher; that is, we cannot give a model that classifies an action as correct or incorrect. This paper proposes a learning procedure using an inductive logic programming (ILP) framework to learn from an inaccurate teacher.

ILP has been intensively studied because it can be applied to complex and structural domains, where the first-order language is necessarily to express objects. Instead of the merit, we can find that the ILP is not very good at the domains of robot learning. Even for simple tasks, such as following a visible

---

* Currently working for Denso corporation

J. Wyatt and J. Demiris (Eds.): EWLR 1999, LNAI 1812, pp. 71–83, 2000.

target, which we describe in this paper, it is not easy to acquire appropriate rules by ILP method. This is an evidence that ILP method should be studied in areas, such as robot learning, which is suffered from the problem of noise and uncertainty. We describe learning in such areas as learning from an inaccutate teacher, which we deal with using ILP framework.

For learning from an inaccurate teacher, we introduce a procedure to collect examples, the procedure which is called an iterative collection of examples (ICE) from experiments with robot learning. ICE repeats the routine of example collection and induction. Example collection is carried out by using previously induced action rules of robots. We will show that action rules induced from examples collected by ICE, incrementally refine a robot's behavior. In the next section, we explain the situation of inaccuracy in our problem. Section 3 introduces the proposed example collection procedure, ICE, and explains its expected effect. Section 4 introduces our experimental environment of robot learning and our ILP problem in this environment. In Section 5, an application of the ICE method to the problem is given, and the results are shown in Section 6. Finally, Section 7 presents some concluding remarks and a discussion of related work.

2   Inaccurate Teachers in Robot Learning

In supervised learning, which includes an ILP framework, a teacher is necessary for classifying cases to be true or otherwise. In some fields of application, however, it is very difficult to provide a teacher, and we can only prepare an inaccurate teacher. Learning for robot control is such a field. It is difficult to classify the actions of a robot at each step.

We define an ideal teacher as an oracle that says yes if and only if a given instance of a target relation is appropriate. We can say a teacher is inaccurate if the teacher is not ideal; that is, the teacher sometimes says yes for an instance even if it is inappropriate, or it says no even if the instance is appropriate.

In robot learning, we face the problem of inaccurate teachers. The reasons why we can only give an inaccurate teacher include:

1. We can get only time-local information and can give only time-local actions at the moment.
2. A global goal (to reach a goal position) cannot be achieved by an action but by a sequence of actions.
3. It is difficult to plan a sequence of actions either for a goal or for a mass of actions, such as avoiding collisions and turning a corner.

From these reasons, we can only give a plausible model to detect preferable actions. A reasonable construction of a teacher for this situation is obtained by combining two teachers.

1. (**A chatty but erroneous teacher**) A model to say yes for appropriate cases and for some inappropriate cases. This can be constructed by weak conditions on a state of a robot at each moment.

Input T : a teacher
$\qquad$ N : the number of examples to be collected
$\qquad$ B : background knowledge

$(E_0^+, E_0^-) :=$ random-collection(T, N) ; $i := 0$
Repeat
$\qquad$ **Induce** a logic program P_i from (E_i^+, E_i^-) with background knowledge B
$\qquad$ $(\Delta E_i^+, \Delta E_i^-) :=$ guided-collection(T, P_i, N)
$\qquad$ $E_{i+1}^+ := E_i^+ \cup \Delta E_i^+$; $E_{i+1}^- := E_i^- \cup \Delta E_i^-$; $i := i+1$
EndRepeat

Fig. 1. Induction algorithm using the iterative collection of examples (ICE).

2. **(A taciturn but reliable teacher)** A model that says no only for critical cases. This can be constructed by watching collisions, for example.

Normally, we can prepare these teachers, because the former teacher can be constructed from a naive understanding of behavior, and the latter can be given based on a global goal.

3   Iterative Collection of Examples

We propose an algorithm, induction by iterative collection of examples (ICE), which is shown in Figure 1. In the algorithm, two example collection procedures, random-collection and guided-collection, are used. They are given in Figure 2.

random-collection collects a specified number of positive and negative examples from the robot's behavior, caused by a random sequence of actions. The actions sequentially change the robot states. This procedure classifies the resulting states into positive and negative examples by using a given teacher. On the other hand, guided-collection collects examples using an induced program of allowable-action. It classifies states caused by actions that are taught by a program.

The induction algorithm by ICE starts with examples collected by random-collection. It induces a program of allowable-action from the examples. Then, it collects examples by guided-collection with the induced program. Collected examples are joined with the previous examples and used to induce a new program. This process is iterated.

We expect that there is an advantage with this learning algorithm, which is explained by the following two effects of iteration loops.

1. **(Effect of counter-examples)** If an induced program is not perfect, it may generate inappropriate actions and may cause inappropriate behavior. Although the chatty but erroneous teacher may not detect this inappropriateness, the taciturn but reliable teacher may detect it. A robot can have an enormous number of situations, and so these kinds of inappropriate actions are observed even after a few iterations of learning. The collected negative

random-collection(T : a teacher; N : int)

> **Put** a robot in a position
> $E^+ := \emptyset; E^- := \emptyset$
> **While** $|E^+| < N$ or $|E^-| < N$ **do**
> > $s :=$ (a sense data at this time)
> > $q_0 :=$ (a robot status at this time)
> > **Choose** a possible action a at random
> > **Let** the robot take the action a
> > $q_1 :=$ (a robot status at this time)
> > **If** T classifies (q_0, q_1) a good action
> > > **then** $E^+ := E^+ \cup \{$allowable-action$(s, a)\}$
> > > **else** $E^- := E^- \cup \{$allowable-action$(s, a)\}$
>
> **EndWhile**
> $E^+ :=$ the first N examples of E^+
> $E^- :=$ the first N examples of E^-
> **Return** (E^+, E^-)

(a)

guided-collection(T : a teacher; P : a program of allowable-action; N : int)

> **Put** a robot in a position
> $E^+ := \emptyset; E^- := \emptyset$
> **While** $|E^+| < N$ or $|E^-| < N$ **do**
> > $s :=$ (a sense data at this time)
> > $q_0 :=$ (a robot status at this time)
> > $a :=$ choose-action(s, P)
> > **Let** the robot take the action a
> > $q_1 :=$ (a robot status at this time)
> > **If** T classifies (q_0, q_1) a good action
> > > **then** $E^+ := E^+ \cup \{$allowable-action$(s, a)\}$
> > > **else** $E^- := E^- \cup \{$allowable-action$(s, a)\}$
>
> **EndWhile**
> $E^+ :=$ the first N examples of E^+
> $E^- :=$ the first N examples of E^-
> **Return** (E^+, E^-)

(b)

Fig. 2. Example collection procedures from (a) random trials and (b) an induced rule.

examples work as counter-examples against the previously induced rules, and refine the rules.

2. **(Effect of accumulated positive examples)** We can expect appropriate behavior from the iteration of induction. If an induced program causes more cases of appropriate behavior than the previous iteration, we can collect more examples of suitable cases. The examples are accumulated as positive examples and used in the induction. If we use a greedy ILP algorithm based on coverage, that is, if a clause (i.e. an if-then rule) that explains a large

part of the examples is chosen first, then clauses that yield appropriate behavior can be expected to be more likely. As a result, this effect gives stable induction of expected clauses in programs.

These two effects are expected to work to induce increasingly good programs. Both effects are related to each other.

4   Problem — Khepera Robots

The architecture of robots A Khepera robot has eight optical sensors and eight infrared sensors. A level of an optical sensor indicates the distance to a source of light, and a combination of levels indicates the direction of the light. The infrared sensors measure distances and direction to objects. A sensor returns an integer value from 0 to 1023. Only the sixteen sensors are a source of information that a Khepera robot can use to decide its actions. We can treat the sensor information as a 16-dimensional vector

$$(o_1, \cdots o_8, i_1, \cdots i_8),$$

where $o_1, \cdots, o_8$ are values of the optical sensors and $i_1, \cdots, i_8$ are values of the infrared sensors. On the other hand, actions that a Khepera can take are to drive right and left wheels from the -1 to 3 levels. Although a real Khepera has the ability to drive in the -10 to 10 level range, we have restricted it. The driving speed of a wheel is in proportion to the value. Negative values mean driving backwards. For a given speed to the two driving wheels a Khepera can control the speed and direction of its motion. Consequently, the motor actions are represented as a 2-dimensional vector

$$(m_l, m_r),$$

where m_l and m_r are the speed levels of the left and right wheels, respectively.

The task for a robot Khepera robots provide a good test-bed for research into robot control. We consider the problem of letting a Khepera robot go from an initial position to a goal position in a maze, which was constructed as shown in Figure 3. A maze is a path from an initial position, denoted by **S** in Figure 3, to a goal, denoted by **G**. A Khepera robot is placed at the initial position and allowed to move using induced rules.

The maze or path is divided into small regions, each of which is divided by a straight line and does not include a bend. The lines dividing the maze are shown by broken lines in Figure 3. At the center of the line separating each region, there is a light that the Khepera robot can use as a guide. The light will be *on* when the Khepera robot enters the region of the light and it will be off when the Khepera leaves. A line, and the light on the line, act as a subgoal for the Khepera robot. The Khepera robot is expected to learn control rules to follow lights and avoid collisions.

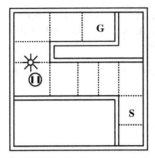

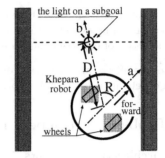

Fig. 3. A maze field.

Fig. 4. Khepera robot and a subgoal.

choose-action(s : sensor information, P : a program of allowable-action)
    Enumerate all possible actions $A = \{a_1, a_2, \cdots, a_n\}$ of robots
    Collect allowable actions $A' = \{a \in A \mid$ allowable-action$(s, a) =$ true wrt $P\}$
    Return choose(A')

choose$_1$(A)
    If $A \neq \emptyset$ **return** an action from A at random
        else return a possible action a at random

choose$_2$(A)
    If $A \neq \emptyset$ **return** an action that equals to the barycenter of A
        else return an action that equals to the barycenter of all of possible actions

Fig. 5. A procedure choose-action to choose an action.

The induction problem For a target robot-learning problem, we consider a relation

$$\text{allowable-action}(s, a),$$

where s is information from a sensor that a robot receives, and a is an action that the robot is allowed to take for good control. This relation indicates a true or false for a given pair of s and a. A procedure to calculate an action that a robot takes, at the moment that information is being received from a sensor, is given in Figure 5. There are choices for these action selection algorithms. Figure 5 contains two algorithms, choose$_1$ and choose$_2$, to switch between in the procedure choose-action. choose$_1$ selects an action from allowable ones at random and choose$_2$ calculates the barycenter; that is, the average motor vector of allowable actions as motor vectors. We use choose$_1$ in the induction processes, and we use two of them in the application processes for comparison.

We did not take the approach of inducing programs to calculate actions from the inputs of sensor information directly, but tried to induce programs of the allowable-action relation. This is because the former approach has difficulty with

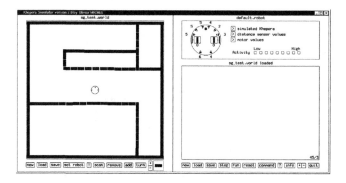

Fig. 6. A snapshot of a Khepera Simulator.

the executability of induced programs. We need special treatments for induction algorithms and for background knowledge, as discussed in [Inuzuka, et al., 1997].

5   Learning Rules of Khepera Robots

In order to learn control rules of Khepera robots in the maze environment by the ICE method, we prepared an inaccurate teacher T.

Teacher T says yes if the condition $(A \lor B) \land \neg C$ is satisfied for an action, and says no otherwise. Conditions A, B, and C are defined as follows:

Condition A : It holds $D - D' \geq 0$,

Condition B : It holds $R - R' \geq 0$, and

Condition C : A collision occurs by the action.

where D and D' are the distances between the center point of Khepera and the light on the subgoal line before and after the action is taken, respectively. The angles R and R' are between Line a (the center line of Khepera), and Line b (the line from Khepera to the light) before and after the action is taken, respectively. These are illustrated in Figure 4.

Condition $A \lor B$ constructs a chatty but erroneous teacher. It is based on the intuitive idea that a Khepera robot should approach a subgoal and turn towards it. This sometimes leads to mistakes because actions obeying this criterion may make Khepera touch the wall, or the shortest path contains actions that are against this criterion. In order to give a weak condition, we give $A \lor B$ but not $A \land B$. The weak condition will most likely contain appropriate conditions but will also contain inappropriate ones. Condition $\neg C$ gives a taciturn but reliable teacher, which is based on the global constraint of avoiding collisions.

To construct the teacher and to collect examples using it, we used free software of a *Khepera Simulator*[Michel, 1996], developed by O. Michel. Figure 6 shows a snapshot of the simulator. It simulates the specification of Khepera

Table 1. Predicates in background knowledge.

predicates	explanations
most-sense-opt(S, N)	Optical sensor N senses the most in S
sense-opt(S, N)	Optical sensor N senses something in S
sense-opt-near(far)(S, N)	Optical sensor N senses more (less) than a level in S
nosense-opt(S, N)	Optical sensor N does not sense in S
most-sense-inf(S, N)	Infrared sensor N senses the most in S
sense-inf(S, N)	Infrared sensor N senses something in S
sense-inf-near(far)(S, N)	Infrared sensor N senses more (less) than a level in S
nosense-inf(S, N)	Infrared sensor N does not sense in S
go-ahead(back)(M)	M drives both wheels forward (back)
move-fast(slow)(M)	M drives wheels fast (slow)
left(right)-turn(M)	M drives right wheel faster (slower) than the left
left(right)-rotate(M)	M drives only right(left) wheel forward
is-forward(back)-sensor(N)	Sensor N is a forward (back) sensor
is-right(left)-forward-sensor(N)	Sensor N is a forward (back) sensor
is-left(right)-side-sensor(N)	Sensor N is a left (right) side sensor
is-opposite-side-sensor(N_1, N_2)	Sensor N_1 is at a mirrored position to sensor N_2
symmetric(N_1, N_2)	Sensor N_1 is at a symmetric position to sensor N_2

robots precisely. We can easily combine a learning method with the simulator. We modified the software to implement the subgoal mechanism and to communicate with Prolog processes, which are necessary for using induced logic programs.

Experimental learning was conducted with the teacher and the Khepera Simulator. At each example collection, 100 positive and 100 negative examples were collected; that is, N in the random-collection and guided-collection is set to 100. For the induction process, we used a FOIL-like ILP system FOIL-I [Inuzuka, et al., 1996,Inuzuka, et al., 1997] with modifications for using an intentional definition of background knowledge. This modified system was used in [Nakano, et al., 1998]. A set of background knowledge was prepared. A part of the background knowledge is shown in Table 1. Background knowledge consists of three groups: knowledge of sensor information; knowledge of motor actions; and knowledge of sensor's positions. They are restricted basic and straightforward predicates.

The detail of the experiment setting is as follows:

1. Collects a sample set S consists of 100 positive and 100 negative examples by random-collection$(T, 100)$, where T is the teacher.
2. Induce a program P of allowable-action using an ILP algorithm from S and background knowledge.
3. Observe behavior of a robot obeying the choose-action procedure and the program P.

4. Collects another 100 positive and 100 negative examples by guided-collection$(T, P, 100)$, and merge them to S. (This step and the previous step can be done simultaneously)
5. Go to 2 and repeat.

To compare our method with a normal setting, we have also done the experiments with normal collection procedure, i.e.:

1. Collects a sample set S consists of 100 positive and 100 negative examples by random-collection$(T, 100)$, where T is the teacher.
2. Induce a program P of allowable-action using an ILP algorithm from S and background knowledge.
3. Observe behavior of a robot obeying the choose-action procedure and the program P.
4. Collects another 100 positive and 100 negative examples by random-collection$(T, 100)$, and merge them to S.
5. Go to 2 and repeat.

6 Results

Figure 7 shows a typical result of the learning process, where the 10 iterations of example collection are carried out. Figure 7(a) shows the numbers of collisions that occurred during the runs of a Khepera robot using each induced program from the ICE method and the random collection method. The numbers are the average of five trials. Figure 7(b) shows the number of collisions that occurred directly from the induced rules. Although a Khepera robot uses induced rules, these sometimes say that no actions are appropriate. Nevertheless, a Khepera robot has to choose an action and so it chooses one at random. Many collisions

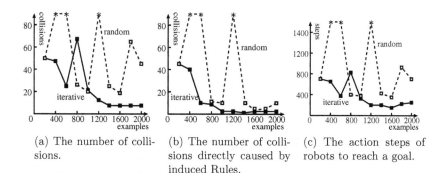

(a) The number of collisions.

(b) The number of collisions directly caused by induced Rules.

(c) The action steps of robots to reach a goal.

Real lines and broken lines show the results of the proposed method and a naive method, respectively. The *'s represents the cases for which a Khepera robot cannot reach a goal.

Fig. 7. Collisions and actions steps of runs by induced rules.

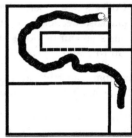
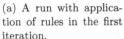

| (a) A run with application of rules in the first iteration. | (b) A run with application of rules in the fourth iteration. | (c) A run with application of rules in the tenth iteration. |

Fig. 8. Traces of runs of a Khepera robot using induced rules in each iteration.

Table 2. Comparison of action selection methods.

	choose$_1$	choose$_2$
collisions occurred	6.8	3.4
steps in a run	240	181

occur for this reason. Figure 7(b) only plots collisions, except those that occurred for this reason. In both graphs, Figure 7(a) and (b), the number of collisions decreased as the number of examples increased. This tendency is clear in the ICE method. The random collection method cannot have a stable effect. Although collisions directly caused by induced rules also decreased in the random collection method, this was not true for the total number of collisions. This may be due to the many examples having the effect of avoiding collisions, but they do not work to induce rules that give actions for all cases. Figure 7(c) shows the steps taken by a Khepera robot in runs with induced rules. This clarifies the effect of rules in efficiency. This graph gives clear evidence that ICE induced better rules. Example traces of runs of a robot using induced rules from the first, fourth, and tenth iterations are shown in Figure 8.

Table 2 summarizes the performance when we used the two different selection methods of actions. choose$_2$, which uses actions calculated as a barycenter of allowable actions, had a better performance.

Figure 9 shows results in a different condition. Here we used a teacher with the following conditions This teacher says yes if the condition $\{(A \wedge B') \vee (A' \wedge B)\} \wedge \neg C$ is satisfied for an action, and says no otherwise. Conditions A, B, and C are the same as the ones defined in Section 5, and Conditions A' and B' are defined as follows:

Condition A' : It holds $D - D' \geq -c \cdot \mathrm{MaxMove}$ and

Condition B' : It holds $R - R' \geq -c \cdot \mathrm{MaxAngle}$,

Case of $c = 0.1$.

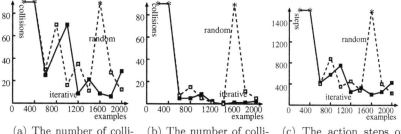

(a) The number of colli-
sions.

(b) The number of colli-
sions directly caused by
induced Rules.

(c) The action steps of
robots to reach a goal.

Case of $c = 0.01$.

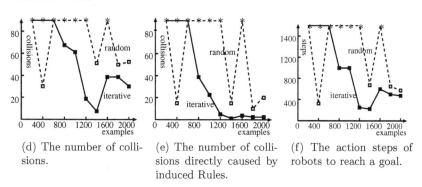

(d) The number of colli-
sions.

(e) The number of colli-
sions directly caused by
induced Rules.

(f) The action steps of
robots to reach a goal.

Real lines and broken lines show the results of the proposed method and a naive
method, respectively. The *'s represents the cases for which a Khepera robot cannot
reach a goal.

Fig. 9. Results from examples with tighter teachers.

where MaxMove is the maximun distance that the robot can move in a step,
MaxAngle the maximum angle that the robot can turn in a step, and c is a
constant in $0 < c$. In the case of $c = \infty$ both Conditions A' and B' are always
held and the teacher is the same as the teacher T in Section 5. The smaller c a
teacher defined, the tigher condition it has. Results of conditions with $c = 0.1$
and $c = 0.01$ is shown in Figure 9. We can observe worse behavior in the case of
$c = 0.1$ and more worse in $c = 0.01$ than the weak teacher (Figure 7), although
the ICE method works much better than the rondom collection procedure.

7   Concluding Remarks

This paper proposed an iterative learning method, ICE, in the ILP framework using the iterative collection of examples method. Examples are collected that are guided by previously induced programs. Negative examples collected using the previous rules are near-misses, which are important negative examples for refining the rules, as emphasized in [Dietterich and Michalski, 1983,Winston, 1975]. Moreover, the positive examples also worked to collect appropriate positive examples, because actions lead to many new situations that are guided by the previously induced rule. As we discussed in Section 3, iteration of induction and collection of examples refine the results effectively. This was evidenced by the experiments using a Khepera robot.

Iterative learning in the field of robot control is a major approach and has been investigated with many ML methods, such as neural networks, genetic algorithms, and reinforcement learning [Franklin, et al., 1996]. The ILP framework, however, has not used iterative learning because a special treatment is needed that refines the induced rules. The proposed method uses iteration, for example collection, and does not need any special mechanism for ILP methods. Direct application of concept learning to robot control is a difficult task. [Kingspor, et al., 1996] tried to learn concepts in robot learning using a hierarchical structure of concepts. Another work related to the control problem includes [Dzeroski, et al., 1995], which tried to induce control rules in a straightforward way. This paper presents the advantages in inducing rules in a difficult situation for robot learning.

Another important contribution of this paper is the presentation of a mechanism for incremental refinement using an ILP framework. This can be applied only for cases that hold the following properties:

1. An inaccurate teacher can be provided with the property explained in Section 2, and
2. Various instances of a target relation can be generated by using induced rules.

For inaccurate teachers, a taciturn but reliable teacher plays important roles for generating counter-examples, in particular. Planning application may have these properties. If we can provide a method for generating instances with induced rules, we may apply the ICE procedure to wider fields.

References

Dietterich and Michalski, 1983. Dietterich, T. G. and Michalski, R. S.: "A comparative review of selected methods for learning from examples", in *Machine Learning: An Artificial Intelligence Approach*, R. S. Michalski, J. Carbonell and T. M. Mitchell (eds.), Palo Alto: Tioga, pp.41–82 (1983).

Dzeroski, et al., 1995. Dzeroski, S., Todorovski, L. and Urbančič, T.: "Handling real numbers in ILP: a step towards better behavioural clones", Proc. 8th European Conf. Machine Learning (ECML'95), LNAI 912, Springer. pp.283–286 (1995).

Franklin, et al., 1996. Franklin, J. A., Mitchell, T. M. and Thrun, S.(eds.), Special Issue on Robot Learning, Machine Learning, Vol.23, No.2/3, (1996).

Inuzuka, et al., 1996. Inuzuka, N., Kamo, M., Ishii, N., Seki, H. and Itoh, H.: "Top-down induction of logic programs from incomplete samples", Proc. 6th Int'l Inductive Logic Programming Workshop (LNAI 1314, Springer-Verlag), pp.265–282 (1996).

Inuzuka, et al., 1997. Inuzuka, N., Seki, H. and Itoh, H.: "Efficient induction of executable logic programs from examples", Proc. Asian Computing Science Conference (LNCS 1345, Springer-Verlag), pp.212–224 (1997).

Kingspor, et al., 1996. Kingspor, V., Morik, K. J. and Dieger, A. D.: "Learning Concepts from Sensor Data of Mobile Robot", Machine Learning, Vol.23, No.2/3, pp.305–332 (1996).

Michel, 1996. Michel, O.: *Khepera Simulator* Package version 2.0: Freeware mobile robot simulator written at the University of Nice Sophia–Antipolis by Olivier Michel. Downloadable from the World Wide Web at `http://wwwi3s.unice.fr/~om/khep-sim.html` (1996).

Nakano, et al., 1998. Nakano, T., Inuzuka, N., Seki, H. and Itoh, H.: "Inducing Shogi Heuristics Using Inductive Logic Programming", Proc. 8th Int'l Conf. Inductive Logic Programming (ILP'98) , LNAI 1446, Springer, pp.155–164 (1998).

Quinlan, 1990. Quinlan, J. R.: "Learning logical definitions from relations", Machine Learning, **5**, pp.239–266 (1990).

Winston, 1975. Winston, P. H.: "Learning Structural Descriptions from Examples", in *The Psychology of Computer Vision*, P. H. Winston (ed.), McGraw-Hill, Inc (1975).

Reinforcement Learning in Situated Agents: Theoretical Problems and Practical Solutions

Mark D. Pendrith[1,2]

[1] School of Computer Science and Engineering,
University of New South Wales, Sydney 2052, Australia
[2] SRI International, Menlo Park CA 94025, USA
pendrith@ieee.org

Abstract. In on-line reinforcement learning, often a large number of estimation parameters (e.g. Q-value estimates for 1-step Q-learning) are maintained and dynamically updated as information comes to hand during the learning process. Excessive variance of these estimators can be problematic, resulting in uneven or unstable learning, or even making effective learning impossible. Estimator variance is usually managed only indirectly, by selecting global learning algorithm parameters (e.g. λ for $TD(\lambda)$ based methods) that are a compromise between an acceptable level of estimator perturbation and other desirable system attributes, such as reduced estimator bias. In this paper, we argue that this approach may not always be adequate, particularly for noisy and non-Markovian domains, and present a direct approach to managing estimator variance, the ccBeta algorithm. Empirical results in an autonomous robotics domain are also presented showing improved performance using the new ccBeta method.

1 Introduction

Many domains of interest in robot learning (and in AI more generally) are too large to be searched exhaustively in reasonable time. One approach has been to employ on-line search techniques, such as *reinforcement learning* (RL) [14].

At first blush, RL presents an excellent conceptual fit for many autonomous robotics applications. However, physically situated learning agents face a number of specific challenges not directly addressed by the Markov decison process/dynamic programming theoretical framework that is conventionally used to analyse RL. In general, situated learning domains will appear to be noisy, non-stationary, and the state of the world will not be fully observable. The learning agent will possibly have to contend with other agents, which may be sympathetic with, benignly indifferent to, or agressively antagonistic towards the agent's goals. There will be real-time constraints that have to be addressed.

All of the above challenges the simplifying assumptions that provide the usual preamble to discussing methods for solving Markov decision processes. In this paper we attempt to address at both a theoretical and practical level a particular problem for which several of the considerations we have listed above have direct import, namely that of managing estimator variance.

J. Wyatt and J. Demiris (Eds.): EWLR 1999, LNAI 1812, pp. 84–102, 2000.

In on-line reinforcement learning, typically a large number of estimation parameters (e.g. Q-value estimates for 1-step Q-learning) are maintained and dynamically updated as information comes to hand during the learning process. Excessive variance of these estimators during the learning process can be problematic, resulting in uneven or unstable learning, or even making effective learning impossible.

Normally, estimator variance is managed only indirectly, by selecting global learning algorithm parameters (e.g. λ for $TD(\lambda)$ based methods) that trade-off the level of estimator perturbation against other system attributes, such as estimator bias or rate of adaptation. In this paper, we give reasons why this approach may sometimes run into problems, particularly for noisy and non-Markovian domains, and present a direct approach to managing estimator variance, the ccBeta algorithm.

1.1   RL as On-Line Dynamic Programming

If an RL method like *1-step Q-learning* (QL) [16] is used to find the optimal policy for a Markov decision process (MDP), the method resembles an asynchronous, on-line form of the dynamic programming *value iteration* method. QL can be viewed as a relaxation method that successively approximates the so-called *Q-values* of the process, the value $Q^\pi(s_t, a_t)$ being the expected value of the return by taking an action a_t from state s_t and following a policy π from that point on.

We note that if an RL agent has access to Q-values for an optimal policy for the system it is controlling, it is easy to act optimally without planning; simply selecting the action from each state with the highest Q-value will suffice. The QL algorithm has been shown to converge, under suitable conditions, to just these Q-values. The algorithm is briefly recounted below.

At each step, QL updates an entry in its table of Q-value estimates according to the following rule (presented here in a "delta rule" form):

$$Q(s_t, a_t) \leftarrow Q(s_t, a_t) + \beta \Delta_t \tag{1}$$

where β is a step-size (or *learning rate*)[1] parameter, and

$$\Delta_t = \mathbf{r}_t^{(1)} - Q(s_t, a_t) \tag{2}$$

where $\mathbf{r}_t^{(1)}$ is the 1-step *corrected truncated return* (CTR):

$$\mathbf{r}_t^{(1)} = r_t + \gamma \max_a Q(s_{t+1}, a) \tag{3}$$

The 1-step CTR is a special case of the n-step CTR. Using Watkins' (1989) notation

$$\mathbf{r}_t^{(n)} = \mathbf{r}_t^{[n]} + \gamma^n \max_a Q(s_{t+n}, a) \tag{4}$$

---

[1] The terms *step-size* and *learning rate* are both used throughout this paper, and are essentially interchangeable.

where $\mathbf{r}_t^{[n]}$ is the simple *uncorrected* n-step truncated return (UTR)

$$\mathbf{r}_t^{[n]} = \sum_{i=0}^{n-1} \gamma^i r_{t+i} \tag{5}$$

As $n \to \infty$, both $\mathbf{r}_t^{[n]}$ and $\mathbf{r}_t^{(n)}$ approach the *infinite horizon* or *actual return*

$$\mathbf{r}_t = \sum_{i=0}^{\infty} \gamma^i r_{t+i} \tag{6}$$

as a limiting case.

We note that if in (2) $\mathbf{r}_t^{(1)}$ were replaced with the actual return, then this would form the update rule for a Monte Carlo estimation procedure. However, rather than waiting for the completed actual return, QL instead employs a "virtual return" that is an estimate of the actual return. This makes the estimation process resemble an on-line value iteration method. One view of the n-step CTR is that it bridges at its two extremes value iteration and Monte Carlo methods. One could also make the observation that such a Monte Carlo method would be an on-line, asynchronous analog of *policy iteration*, another important dynamic programming technique.

The central conceptual importance of the CTR to RL techniques is that virtually all RL algorithms estimate the value function of the state/action pairs of the system using either single or multi-step CTRs directly, as in the case of QL or the C-Trace algorithm [8], or as returns that are equivalent to weighted sums of varying length n-step CTRs, such as the $TD(\lambda)$ return [15,16].

2   CTR Bias and Variance

For RL in Markovian domains, the choice of length of CTR is usually viewed as a trade-off between bias and variance of the sample returns to the estimation parameters, and hence of the estimation parameters themselves (e.g. Watkins 1989).

The idea is that shorter CTRs should exhibit less variance but more bias than longer CTRs. The increased bias will be due to the increased weight in the return values of estimators that will, in general, be inaccurate while learning is still taking place. The expected reduction in estimator variance is due to the fact that for a UTR the variance of the return will be strictly non-decreasing as n increases.

Applying this reasoning uncritically to CTRs is problematic, however. In the case of CTRs, we note that initial estimator inaccuracies that are responsible for return bias may also result in high return variance in the early stages of learning. Thus, in the early stages of learning, shorter CTRs may actually result in the worst of both worlds – high bias *and* high variance.

By way of illustration, consider the simple MDP depicted in Figure 1. The expected variance of the sampled returns for action 0 from state A will be arbitrarily high depending upon the difference between the initial estimator values

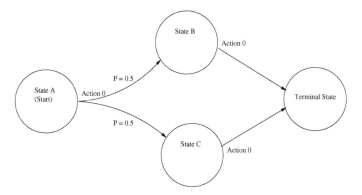

Fig. 1. A 4-state/1 action MDP. State A is the starting state, states B and C are equiprobable successor states after taking action 0. Actions from B and C immediately lead to termination. The immediate reward at each step is zero. If 1-step CTRs are used, the variance as well as bias of the estimator returns for State A/action 0 depends on the difference of the initial estimator values for states B and C. On the other hand, if CTRs of length 2 or greater are used, the estimator returns will be unbiased and have zero variance.

for actions from states B and C. In this case, the estimator for $\langle A, 0 \rangle$ would experience both high bias and high variance if 1-step CTRs were to be used. On the other hand, using CTRs of length 2 or greater would result in unbiased estimator returns with zero variance at all stages of learning for this MDP.

In general, as estimators globally converge to their correct values, the variance of an n-step CTR for an MDP will become dominated by the variance in the terms comprising the UTR component of the return value, and so the relation

$$i < j \Rightarrow var[\mathbf{r}_t^{(i)}] \leq var[\mathbf{r}_t^{(j)}] \tag{7}$$

will be true in the limit. However, the point we wish to make here is that a bias/variance tradeoff involving n for CTRs is not as clear cut as may be often assumed, particularly in the early stages of learning, or at any stage of learning if the domain is noisy,[2] even if the domain is Markovian.

2.1 CTR Bias and Variance in NMDPs

Perhaps more importantly, if the domain is not Markovian, then the relation expressed in (7) is not guaranteed to hold for *any* stage of the learning. To demonstrate this possibly surprising fact, we consider the simple 3-state non-Markov Decision Process (NMDP) depicted in Figure 2.

---

[2] The argument here is that for noisy domains, estimator bias is continually being reintroduced, taking the process "backwards" towards the conditions of early learning.

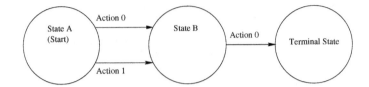

Fig. 2. A 3-state NMDP, with two available actions from starting state A, and one available from the successor state B. The action from state B immediately leads to termination and a reward; the decision process is non-Markov because the reward depends on what action was previously selected from state A.

For this NMDP, the expected immediate reward from taking the action 0 in state B depends upon which action was taken from state A. Suppose the (deterministic) reward function ρ is as follows:

$$\rho(A, 0) = 0$$
$$\rho(A, 1) = 0$$
$$\rho(B, 0) = 0 \quad \text{(if the action from state A was 0)}$$
$$\rho(B, 0) = 1 \quad \text{(if the action from state A was 1)}$$

If Monte Carlo returns are used (or, equivalently in this case, n-step CTRs where $n > 1$), the estimator returns for state/action pairs $\langle A, 0 \rangle$ and $\langle A, 1 \rangle$ will exhibit zero variance at all stages of learning, and the corresponding estimators should rapidly converge to their correct values of 0 and 1 respectively.

On the other hand, if 1-step CTRs are used, the variance of the $\langle A, 0 \rangle$ and $\langle A, 1 \rangle$ estimators will be non-zero while the variance of the estimator for $\langle B, 0 \rangle$ is non-zero. The estimator for $\langle B, 0 \rangle$ will exhibit non-zero variance as long as both actions continue to be tried from state A, which would normally be for all stages of learning for the sake of active exploration. Finally, note that the variance for $\langle B, 0 \rangle$ will the same in this case for all n-step CTRs $n \geq 1$. Hence, the overall estimator variance for this NMDP is *strictly greater at all stages of learning* for 1-step CTRs than for any n-step CTRs $n > 1$.

In previously published work studying RL in noisy and non-Markovian domains [8], excessive estimator variance appeared to be causing problems for 1-step QL in domains where using Monte Carlo style returns improved matters. These unexpected experimental results did not (and still do not) fit well with the "folk wisdom" concerning estimator bias and variance in RL. We present these analyses firstly as a tentative partial explanation for the unexpected results in these experiments.

Secondly, the foregoing analysis is intended to provide some theoretical motivation for an entirely different approach to managing estimator variance in RL, in which attention is shifted away from CTR length and is instead focused on the step-size parameter β. In the next part of this paper, we discuss a new algorithm (perhaps more accurately a family of algorithms) we call ccBeta, which results from taking such an approach.

3 β: Variance versus Adaptability

In RL, finding a good value for the step-size parameter β for a particular algorithm for a particular domain is usually in a trial-and-error process for the experimenter, which can be time consuming. The resulting choice is usually a trade-off between fast adaptation (large β) and low estimator variance (small β). In RL, and in adaptive parameter estimation systems generally, there emerges a natural tension between the issues of convergence and adaptability.

Stochastic convergence theory [7,9] suggests that a *reducing* β series (such as $\beta_i = 1/i$) that satisfies the Robbins and Monro criteria

$$\sum_{i=1}^{\infty} \beta_i = \infty, \quad \text{and} \quad \sum_{i=1}^{\infty} \beta_i^2 < \infty \tag{8}$$

may be used to adjust an estimator's value for successive returns; this will guarantee in-limit convergence under suitable conditions. However, this is in general not a suitable strategy for use in non-stationary environments i.e. environments in which the schedule of payoffs may vary over time. While convergence properties for stationary environments are good using this technique, re-adaptation to changing environmental conditions can be far too slow.

In practice, a constant value for the β series is usually chosen. This has the advantage of being constantly sensitive to environment changes, but has poorer convergence properties, particularly in noisy or stochastic environments. It appears that the relatively poor convergence properties of a constant β series can lead to instabilities in learning in some situations, making an effective trade-off between learning rate and variance difficult.

A method for automatically varying the step-size β parameter by a simple on-line statistical analysis of the estimate error is presented here. The resulting β series will be neither constant nor strictly decreasing, but will vary as conditions indicate.

3.1   The ccBeta Algorithm

At each update step for the parameter estimate, we assume we are using a "delta-rule" or on-line LMS style update rule along the lines of

$$\Delta_i \leftarrow z_i - Q_{i-1}$$
$$Q_i \leftarrow Q_{i-1} + \beta_i \Delta_i$$

where z_i is the ith returned value in the series we are trying to estimate and β_i is the ith value of the step-size schedule series used. Q_i is the ith estimate of this series, and Δ_i is the ith value of the error series.

The idea behind ccBeta is quite straightforward. If the series of estimate errors for a parameter is positively auto-correlated, this indicates a persistent over- or under-estimation of the underlying value to be estimated is occurring, and suggests β should be increased to facilitate rapid adaptation. If, on the other hand, the estimate errors are serially uncorrelated, then this may be taken as an indication that there is no systematic error occurring, and β can be safely decreased to minimise variance while these conditions exist.

So, for each parameter we are trying to estimate, we keep a separate set of autocorrelation statistics for its error series as follows, where cc_i is derived as an exponentially weighted autocorrelation coefficient:

$$sum\_square\_err_i \leftarrow K.sum\_square\_err_{i-1} + \Delta_i^2 \tag{9}$$

$$sum\_product_i \leftarrow K.sum\_product_{i-1} + \Delta_i.\Delta_{i-1} \tag{10}$$

$$cc_i \leftarrow \frac{sum\_product_i}{\sqrt{sum\_square\_err_i.sum\_square\_err_{i-1}}} \tag{11}$$

At the start of learning, the sum_square_err and $sum\_product$ variables are initialised to zero; but this potentially leads to a divide-by-zero problem on the RHS of (11). We explicitly check for this situation, and when detected, cc_i is set to 1.[3]

We note that if in (9) and (10) the exponential decay parameter $K \in [0, 1]$ is less than 1, two desirable properties emerge: firstly, the values of the sum_square_err and $sum\_product$ series are finitely bounded, and secondly the correlation coefficients are biased with respect to recency. While the first property is convenient for practical implementation considerations with regard to possible floating point representation overflow conditions etc., the second property is essential for effective adaptive behaviour in non-stationary environments. Setting K to a value of 0.9 has been found to be effective in all domains tested so far; experimentally this does not seem to be a particularly sensitive parameter.

It is also possible to derive an autocorrelation coefficient not of the error series directly, but instead of the sign of the error series, i.e. replacing the Δ_i and Δ_{i-1} terms in (9) and (10) with $sgn(\Delta_i)$ and $sgn(\Delta_{i-1})$. This variant may prove to be generally more robust in very noisy environments.

In such a situation an error series may be so noisy that, even if the error signs are consistent, a good linear regression is not possible, and so β will be small even when there is evidence of persistent over- or under-estimation. This approach proved to be successful when applied to the extremely noisy real robot domain described in the next section. Based on our results to date, this version could be recommended as a good "general purpose" version of ccBeta.

Once an autocorrelation coefficient is derived, β_i is set as follows:

> if $(cc_i > 0)$
> $\quad \beta_i \leftarrow cc_i * MAX\_BETA$
> else $\quad \beta_i \leftarrow 0$

> if $(\beta_i < MIN\_BETA)$
> $\quad \beta_i \leftarrow MIN\_BETA$

---

[3] This may seem arbitrary, but the reasoning is simply that if you have exactly one sample from a population to work with, the best estimate you can make for the mean of that population is the value of that sample.

First, we note that in the above pseudo-code, negative and zero auto-correlations are treated the same for the purposes of weighting β_i. A strongly negative auto-correlation indicates alternating error signs, suggesting fluctuations around a mean value. Variance minimisation is also desirable in this situation, motivating a small β_i.

On the other hand, a strongly positive cc_i results from a series of estimation errors of the same sign, indicating a persistent over- or under-estimation, suggesting a large β_i is appropriate to rapidly adapt to changed conditions.

Setting the scaling parameters MIN_BETA to 0.01 and MAX_BETA to 1.0 has been found to be effective, and these values are used in the experiments that follow. Although values of 0 and 1 respectively might be more "natural", as this corresponds to the automatic scaling of the correlation coefficient, in practice a small non-zero MIN_BETA value was observed to have the effect of making an estimate series less discontinuous (although whether this offers any real advantages has not been fully determined at this stage.)

Finally, we note that *prima facie* it would be reasonable to use cc_i^2 rather than cc_i to weight β_t. Arguably, cc_i^2 is the more natural choice, since from statistical theory it is the square of the correlation coefficient that indicates the proportion of the variance that can be attributed to the change in a correlated variable's value.

Both the cc_i and cc_i^2 variations have been tried, and while in simulations marginally better results both in terms of total square error (TSE) and variance were obtained by using cc_i^2, a corresponding practical advantage was not evident when applied to the robot experiments.

4   Related Work

Darken, Chang and Moody [4] described a heuristic "search then converge" reducing learning rate schedule designed for use with backpropagation networks. The essential idea is that the learning rate remains almost constant at the beginning of the series ("search mode") but after a certain point (determined by parameter choice) starts reducing roughly in inverse proportion to time step t ("converge mode").[4] Thus, we would expect the adaptation properties of such a schedule would resemble that of a fixed rate schedule while in "search mode" (fast adaptation but high estimator variance), and a reducing schedule when in "converge mode" (poor adaptation but low estimator variance).

According to Jacobs [5], Kesten [6] first suggested that an alternating error sign could be used as an indication that a step-size parameter should be reduced in the context of a stochastic approximation method, and that Saridis [10] extended the idea to both increasing and decreasing the learning rate (i.e. if successive errors are of the same sign, the step-size should be incremented).

---

[4]   Bradtke and Barto [2] applied Darken et al.'s learning rate scheduler to an online temporal difference reinforcement learning algorithm; they conjectured that the way they did this would indirectly satisfy the Robbins and Monro criteria (refer to Equation 8) for convergence of their estimators.

Jacobs himself proposes the Delta-Bar-Delta learning rate update rule for use in the context of multi-layer nonlinear backpropagation networks. This is a refinement of earlier work by Barto & Sutton [1], and Sutton [12]. Jacobs contrasts the properties of his proposed update rule with some of the shortcomings of previously proposed methods. Jacobs also proposes four general heuristics applicable to a step-size adjustment method:

- Every system parameter should have its own learning rate
- Every learning rate should be allowed to vary over time
- When the derivative of a parameter possesses the same sign for several consecutive time steps, the learning rate for that parameter should be increased
- When the sign of the derivative of a parameter alternates for several consecutive time steps, the learning rate for that parameter should be decreased

We note that the ccBeta method proposed in this paper satisfies these principles, even though Jacobs was primarily concerned with improving rates of convergence over steepest descent techniques in the context of backpropagation networks. Further, we might claim that ccBeta uses a more statistically principled approach in adjusting the step-size by directly calculating the 1-step autocorrelation of the error signal time series; the heuristics proposed for Delta-Bar-Delta method are plausible but without a clear theoretical underpinning. Jacobs himself claims no special status for the implementation of these heuristics he proposes, concluding his paper with the comment "Clearly there are other implementations of the heuristics worth studying."

Sutton [13] later proposed an incremental version of the Delta-Bar-Delta rule (IDBD) for use in a linear function approximator. Like Jacobs and Darken et al., he analyses his method in terms of gradient descent, which is natural in the context of stochastic function approximation when all system weights are generally updated in parallel. In the context of on-line RL, as we discuss in this paper, this analysis does not apply directly, however. To point to one obvious difference, typically only a subset of state or state/action pair value estimators are updated upon each time step.

Nonetheless, all of the basic intuitions that motivate the learning rate adaptation techniques for stochastic approximation methods apply equally well to the situation where we are interested in updating Q-value or state value estimates in an on-line RL setting. In the experiments that are described in the next section, we test these intuitions directly.

5   Experimental Results

In all the experiments desribed in this section we use the same variant of ccBeta; this has a K parameter of 0.9, and uses $sgn(\Delta)$ normalisation to calculate cc_i. For weighting β_i, we use cc_i rather than cc_i^2, and have MIN_BETA and MAX_BETA set to 0.01 and 1.0 respectively.

5.1   Simulation Experiment 1

In this section, we describe some simulation studies comparing the variance and re-adaptation sensitivity characteristics of the ccBeta method for generating a β step-size series against standard regimes of fixed β and reducing β, where $\beta_i = 1/i$.

The learning system for these experiments was a single-unit on-line LMS estimator which was set up to track an input signal for 10,000 time steps. In the first experiment, the signal was stochastic but with stationary mean: a zero function perturbed by uniform random noise in the range $[-0.25, +0.25]$. The purpose of this experiment was to assess asymptotic convergence properties, in particular estimator error and variance.

As can be seen from Table 1, the reducing beta schedule of $\beta_i = 1/i$ was superior to fixed beta and ccBeta in terms of both total square error (TSE) and estimator variance for this experiment. As we would expect, variance (normalised to standard deviation in the tabled results) increased directly with the magnitude of beta for the fixed beta series. In this experiment ccBeta performed at a level between fixed beta set at 0.1 and 0.2.

Table 1. Results for simulation experiment 1.

Simulation experiment 1 (noisy zero fn)		
	T.S.E.	Std. Dev.
ccBeta	17.95	0.042
beta = 0.1	10.42	0.032
beta = 0.2	22.30	0.047
beta = 0.3	35.72	0.060
beta = 0.4	50.85	0.071
reducing beta	0.17	0.004

5.2   Simulation Experiment 2

In the second experiment, a non-stationary stochastic signal was used to assess re-adaptation performance. The signal was identical to that used in the first experiment, except that after the first 400 time steps the mean changed from 0 to 1.0, resulting in a noisy step function. The adaptation response for the various beta series over 50 time steps around the time of the change in mean are plotted in Figures 3 and 4.

To get an index of responsiveness to the changed mean using the different beta series, we have measured the number of time steps from the time the mean level was changed to the time the estimator values first crossed the new mean level, i.e. when the estimator first reaches a value ≥ 1.

As can be seen from Table 2, the reducing beta schedule of $\beta_i = 1/i$ was far worse than for either fixed beta or ccBeta in terms of TSE and re-adaptation performance ("Steps to Crossing", column 2). Indeed, the extreme sluggishness of the reducing beta series was such that the estimator level had risen to only about 0.95 after a further 9,560 time steps past the end of the time step window shown in Figure 3. The relatively very high TSE for the reducing beta series was also almost entirely due to this very long re-adaptation time. The inherent unsuitability of such a regime for learning in a non-stationary environment is clearly illustrated in this experiment. Despite having "nice" theoretical properties, it represents an impractical extreme in the choice between good in-limit convergence properties and adaptability.

Table 2. Results for simulation experiment 2.

Simulation experiment 2 (noisy step fn)		
	T.S.E.	Steps to Crossing
ccBeta	21.60	10
beta = 0.1	16.19	35
beta = 0.2	25.84	15
beta = 0.3	38.56	9
beta = 0.4	53.35	8
reducing beta	395.73	>9,600

In the Figure 3 plots, the trade-off of responsiveness versus estimator variance for a fixed beta series is clearly visible. We note however that the re-adaptation response curve for ccBeta (Figure 4) resembles that of the higher values of fixed beta, while its TSE (Table 2) corresponds to lower values, which gives some indication the algorithm is working as intended.

5.3   Experiment 3: Learning to Walk

For the next set of experiments we have scaled up to a real robot-learning problem: the gait coordination of a six-legged insectoid walking robot.

The robot (affectionately referred to as "Stumpy" in the UNSW AI lab) is faced with the problem of learning to walk with a forward motion, minimising backward and lateral movements. In this domain, ccBeta was compared to using hand tuned fixed beta step-size constants for two different versions of the C-Trace RL algorithm.

Stumpy is, by robotics standards, built to an inexpensive design. Each leg has two degrees of freedom, powered by two hobbyist servo motors. A 68HC11 Miniboard converts instructions from a 486 PC sent via a RS-232-C serial port connection into the pulses that fire the servo motors. The primary motion sensor is a cradle-mounted PC mouse dragged along behind the robot. This provides for very noisy sensory input, as might be appreciated. The quality of the signal has been found to vary quite markedly with the surface the robot is traversing.

As well as being noisy, this domain was non-Markovian by virtue of the compact but coarse discretized state-space representation. This compact repre-

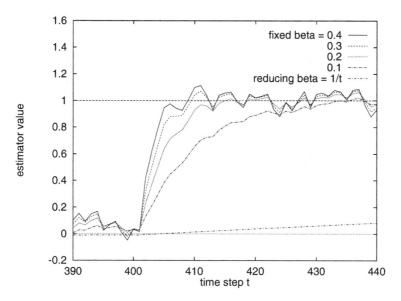

Fig. 3. Fixed beta and reducing beta step response plots (simulation experiment 2).

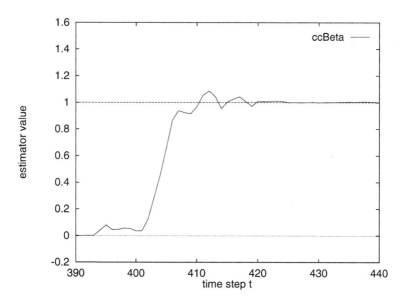

Fig. 4. ccBeta step response plot (simulation experiment 2).

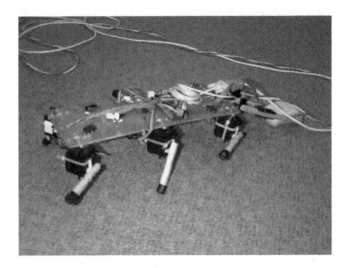

Fig. 5. The robot learning to walk.

sentation[5] meant learning was fast, but favoured an RL algorithm that did not rely heavily on the Markov assumption; in earlier work [8] C-Trace had been shown to be well-suited for this domain.

The robot was given a set of primitive "reflexes" in the spirit of Brooks' "subsumption" architecture [3]. A leg that is triggered will incrementally move to lift up if on the ground and forward if already lifted. A leg that does not receive activation will tend to drop down if lifted and backwards if already on the ground. In this way a basic stepping motion was encoded in the robots "reflexes".

The legs were grouped to move in two groups of three to form two tripods (Figure 6). The learning problem was to discover an efficient walking policy by triggering or not triggering each of the tripod groups from each state. Thus the action set to choose from in each discretized state consisted of four possibilities:

- Trigger both groups of legs
- Trigger group A only
- Trigger group B only
- Do not trigger either group

The robot received positive reinforcement for forward motion as detected by the PC mouse, and negative reinforcement for backward and lateral movements.

Although quite a restricted learning problem, interesting non-trivial behaviours and strategies have been seen to emerge.

---

[5] 1024 "boxes" in 6 dimensions: alpha and beta motor positions for each leg group (4 continuous variables each discretized into 4 ranges), plus 2 Boolean variables indicating the triggered or untriggered state of each leg group at the last control action.

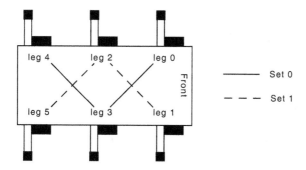

Fig. 6. Grouping the legs into two tripods.

5.4   The RL Algorithms

As mentioned earlier, C-Trace is an RL algorithm that uses multi-step CTRs to estimate the state/action value function. While one C-Trace variant, *every-visit C-Trace* (Figure 7), has been described in earlier work in application to this domain [8], the other, *first-visit C-Trace* (Figure 7), is a variant that has not been previously described.

As mentioned earlier, C-Trace is an RL algorithm that uses n-step CTRs to estimate the state/action Q-value function. The length n of CTRs vary, a return truncation/correction occurring on the time steps when a non-policy action is selected for the sake of active exploration. The idea is to maximise the average length of n to minimise bias and to speed learning, while retaining the QL property of "experimentation insensitivity", which means that the effects of active exploration do not affect the value function being learnt (at least not for Markov domains) [8].

It is easiest to understand the difference between an every-visit and first-visit C-Trace in terms of the difference between an every-visit and a first-visit Monte Carlo algorithm. Briefly, a first-visit Monte Carlo algorithm will selectively ignore some returns for the purposes of learning in order to get a more truly independent sample set. In [11], first-visit versus every-visit returns are discussed in conjunction with a new "first-visit" version of the $TD(\lambda)$ algorithm, and significant improvements in learning performance are reported for a variety of domains using the new algorithm.

For these reasons, a first-visit version of C-Trace seemed likely to be particularly well-suited for a ccBeta implementation, as the "purer" sampling methodology for the returns should (in theory) enhance the sensitivity of the on-line statistical tests.

5.5   Discussion of Results

The average forward walking speed over the first hour of learning for two versions of C-Trace using both ccBeta and fixed step-size parameters are presented in the plots in Figures 9 and 10.

```
for all s ∈ S, a ∈ A do
        VisitCount_sa ← 0                              ( initialise traces )
endfor

while not terminal state do

        get current state s
        select action a                                ( stochastic action selection )

        if non-policy action selected then
                for all i ∈ S, j ∈ A such that VisitCount_ij > 0 do
                        k ← GlobalClock − StepCount_ij
                        c ← max_m Q_im
                        ρ_ij ← ρ_ij + γ^k c              ( truncated return correction )
                        n ← VisitCount_ij
                        Sum_ij ← (1 − β)Sum_ij + x_ij ρ_ij
                        Q_ij ← (1 − β)^n Q_ij + Sum_ij   ( apply CTR update )
                        VisitCount_ij ← 0               ( zero traces )
                endfor
        endif

        if VisitCount_sa = 0 then                       ( first visit )
                Sum_sa ← 0
                x_sa ← β
        else                                            ( subsequent visits )
                Sum_sa ← (1 − β)Sum_sa + x_sa ρ_sa
                k ← GlobalClock − StepCount_sa           ( steps since last visit )
                x_sa ← (1 − β)x_sa γ^k + β
        endif

        ρ_sa ← 0
        VisitCount_sa ← VisitCount_sa + 1
        StepCount_sa ← GlobalClock                       ( "time-stamp" visit )

        take action a

        if reinforcement signal r received then
                for all (i, j) such that VisitCount_ij > 0 do
                        k ← GlobalClock − StepCount_ij
                        ρ_ij ← ρ_ij + γ^k r
                endfor
        endif

        GlobalClock ← GlobalClock + 1
endwhile
for all s ∈ S, a ∈ A such that VisitCount_sa > 0 do      ( terminal state reached )
        n ← VisitCount_sa
        Sum_sa ← (1 − β)Sum_sa + x_sa ρ_sa
        Q_sa ← (1 − β)^n Q_sa + Sum_sa                   ( terminal update rule )
endfor
```

Fig. 7. Pseudo-code for every-visit C-Trace. S and A denote the state set and the action set respectively. The variables Q, Sum, x, ρ, $StepCount$ and $VisitCount$ are kept separately for each state/action pair. The subscripts identify to which state/action pair the variable belongs; e.g. Q_{ij} is the Q-value estimate for state i and action j.

In the case of every-visit C-Trace (Figure 9), we notice immediately that the learning performance is much more stable using ccBeta than with the fixed

```
for all s ∈ S, a ∈ A do
        VisitCount_sa ← 0                          ( initialise traces )
endfor

while not terminal state do

        get current state s
        select action a                            ( stochastic action selection )

        if non-policy action selected then
                for all i ∈ S, j ∈ A such that VisitCount_ij > 0 do
                        k ← GlobalClock − StepCount_ij
                        c ← max_m Q_im
                        ρ_ij ← ρ_ij + γ^k c         ( truncated return correction )
                        Q_ij ← (1 − β)Q_ij + βρ_ij  ( apply CTR update )
                        VisitCount_ij ← 0           ( zero traces )
                endfor
        endif

        if VisitCount_sa = 0 then                   ( first visit )
                ρ_sa ← 0
                VisitCount_sa ← 1
                StepCount_sa ← GlobalClock          ( "time-stamp" visit )
        endif

        take action a

        if reinforcement signal r received then
                for all i ∈ S, j ∈ A such that VisitCount_ij > 0 do
                        k ← GlobalClock − StepCount_ij
                        ρ_ij ← ρ_ij + γ^k r
                endfor
        endif

        GlobalClock ← GlobalClock + 1
endwhile

for all s ∈ S, a ∈ A such that VisitCount_sa > 0 do   ( terminal state reached )
        Q_sa ← (1 − β)Q_sa + βρ_sa                     ( terminal update rule )
endfor
```

Fig. 8. Pseudo-code for first-visit C-Trace. S and A denote the state set and the action set respectively. The variables Q, ρ, $StepCount$ and $VisitCount$ are kept separately for each state/action pair. The subscripts identify to which state/action pair the variable belongs; e.g. Q_{ij} is the Q-value estimate for state i and action j.

beta series. This shows up as obviously reduced variance in the average forward speed; significantly, the overall learning rate doesn't seem to have been adversely affected, which is encouraging. It would not be unreasonable to expect some trade-off between learning stability and raw learning rate, but such a trade-off is not apparent in these results.

Interestingly, the effects of estimator variance seem to manifest themselves in a subtly different way in the first-visit C-Trace experiments (Figure 10). We notice that first-visit C-Trace even without ccBeta seems to have had a marked effect on reducing the step-to-step variance in performance as seen in every-visit

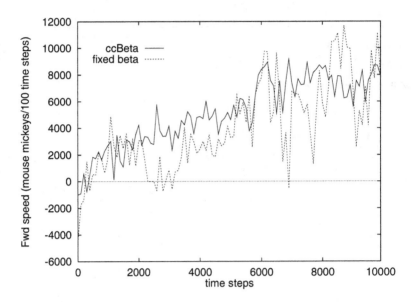

Fig. 9. The robot experiment using RL algorithm 1 (every-visit C-Trace).

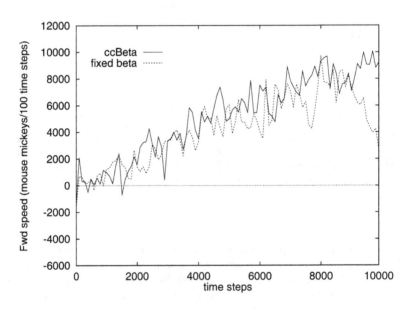

Fig. 10. The robot experiment using RL algorithm 2 (first-visit C-Trace).

C-Trace. This is very interesting in itself, and calls for further theoretical and experimental investigation.[6]

However, we also notice that at around time step 8000 in the first-visit fixed-beta plot there is the start of a sharp decline in performance, where the learning seems to have become suddenly unstable.

These results were averaged over multiple runs ($n = 5$) for each plot, so this appears to be a real effect. If so, it is conspicuous by its absence in the first-visit ccBeta plot: ccBeta would appear to have effectively rectified the problem.

Overall, the combination of first-visit C-Trace and ccBeta seems to be the winning combination for these experiments, which is encouragingly in agreement with prediction.

6 Conclusions

Excessive estimator variance during on-line learning can be a problem, resulting in learning instabilities of the sort we have seen in the experiments described here.

In RL, estimator variance is traditionally dealt with only indirectly via the general process of tuning various learning parameters, which can be a time consuming trial-and-error process. Additionally, theoretical analysis presented here indicate that some of the pervasive ideas regarding the trade-offs involved in the tuning process need to be critically examined. In particular, the relationship between CTR length and estimator variance needs reassessment, particularly in the case of learning in a non-Markov domain.

The ccBeta algorithm has been presented as a practical example of an alternative approach to managing estimator variance in RL. This algorithm has been designed to actively minimise estimator variance while avoiding the degradation in re-adaptation response times characteristic of passive methods. ccBeta has been shown to perform well both in simulation and in real-world learning domains.

A possible advantage of this algorithm is that since it is not particularly closely tied in its design or assumptions to RL algorithms, Markov or otherwise, it may turn out be usable with a fairly broad class of on-line search methods. Basically, any method that uses some form of "delta rule" for parameter updates might potentially benefit from using a ccBeta-style approach to managing on-line estimator variance.

---

[6] At this point, we will make the following brief observations on this effect: a) The reduction in variance theoretically makes sense inasmuch as the variance of the sum of several random variables is equal to the sum of the variances of the variables if the variables are not correlated, but will be greater than this if they are positively correlated. With every-visit returns, the positive correlation between returns is what prevents them being statistically independent. b) This raises the interesting possibility that the observed improved performance of "replacing traces" owes as much if not more to a reduction in estimator variance than to reduced estimator bias, which is the explanation proposed by [11].

Acknowledgements

The author is indebted to Paul Wong and Malcolm Ryan for their assistance in performing the robot experiments described in this paper. Peter Neilson

References

1. A. G. Barto and R. S. Sutton. Goal seeking components for adaptive intelligence: An initial assessment. Technical Report AFWAL-TR-81-1070, Air Force Wright Aeronautical Laboratories/Avionics Laboratory, Wright-Patterson AFB, Ohio, 1981.
2. S. J. Bradtke and A. G. Barto. Linear least-squares algorithms for Temporal Difference learning. *Machine Learning*, 22(1–3), 1996.
3. Rodney Brooks. Intelligence without reason. In *Proceedings of the 12th International Joint Conference on Artificial Intelligence*, pages 569–595, 1991.
4. C. Darken, C. Chang, and J. Moody. Learning rate schedules for faster stochastic gradient search. In *Neural networks for signal processing 2 – Proceedings of the 1992 IEEE workshop*. IEEE Press, 1992.
5. Robert A. Jacobs. Increased rates of convergence through learning rate adaptation. *Neural Networks*, 1:295–307, 1988.
6. H. Kesten. Accelerated stochastic approximation. *Annals of Mathematical Statistics*, 29:41–59, 1958.
7. H.J. Kushner and D.S. Clark. *Stochastic approximation methods for constrained and unconstrained systems*. New York: Springer-Verlag, 1978.
8. M.D. Pendrith and M.R.K. Ryan. Actual return reinforcement learning versus Temporal Differences: Some theoretical and experimental results. In L.Saitta, editor, *Machine Learning: Proc. of the Thirteenth Int. Conf.* Morgan Kaufmann, 1996.
9. H. Robbins and S. Monro. A stochastic approximation method. *Annals of Mathematical Statistics*, 22:400–407, 1951.
10. G. N. Saridis. Learning applied to successive approximation algorithms. *IEEE Transactions on Systems Science and Cybernetics*, SSC-6:97–103, 1970.
11. S.P. Singh and R.S. Sutton. Reinforcement learning with replacing eligibility traces. *Machine Learning*, 22:123–158, 1996.
12. R. S. Sutton. Two problems with backpropagation and other steepest-descent learning procedures for networks. In *Proceedings of the Eight Annual Conference of the Cognitive Science Society*, pages 823–831, 1986.
13. R. S. Sutton. Adapting bias by gradient descent: An incremental version of Delta-Bar-Delta. In *Proceeding of the Tenth National Conference on Artificial Intelligence*, pages 171–176. MIT Press, 1992.
14. R. S. Sutton and A. G. Barto. *Reinforcement Learning: An Introduction*. MIT Press, 1998.
15. R.S. Sutton. Learning to predict by the methods of temporal difference. *Machine Learning*, 3:9–44, 1988.
16. C.J.C.H. Watkins. *Learning from Delayed Rewards*. Ph.D. Thesis, King's College, Cambridge, 1989.

A Planning Map for Mobile Robots: Speed Control and Paths Finding in a Changing Environment

Mathias Quoy[1], Philippe Gaussier[1], Sacha Leprêtre[1],
Arnaud Revel[1], and Jean-Paul Banquet[2]

[1] Neurocybernetics team
ETIS - Université de Cergy-Pontoise - ENSEA
6, Av du Ponceau, 95014 Cergy Pontoise Cedex, FRANCE
quoy@u-cergy.fr,
http://www-etis.ensea.fr
[2] Institut Neurosciences et Modélisation, Paris VI

Abstract. We present here a neural model for mobile robot action selection and trajectories planning. It is based on the elaboration of a
"cognitive map". This cognitive map builds up a graph linking together
reachable places. We first demonstrate that this map may be used for
the control of the robot speed assuring a convergence to the goal. We
show afterwards that this model enables to select between different goals
in a static environment and finally in a changing environment.

1 Introduction

One of our research interests is to propose architectures for controlling mobile
robots. These architectures are mainly constrained by the facts that we
want the representations to be "grounded" on the sensor data (*construc-
tivist approach*) and that we want an analogical computation at each stage.
These requirements particularly fit in the neural network (N.N.) framework.
Furthermore, the N.N. approach enables to take inspiration from neurobiological and psychological results, as well for the architectures as for the
learning processes (see [Baloch and Waxman, 1991], [Millan and Torras, 1992],
[Verschure and Pfeifer, 1992], [Gaussier and Zrehen, 1994] for other works on
this issue). We also stress that our robot has its own internal "motivations"
and that the whole learning process is under their influence. This is very
important because our system has to "behave" autonomously in an a priori
unknown environment (*animat approach* [Meyer and Wilson, 1991]). So we will
call our robot an "animat". Our control architecture is mainly composed of
two parts. The first one enables the animat to locate itself in the environment
[Gaussier et al., 1997a]. We will not discuss about this part in this paper. The
second one builds a graph linking successively reached places ("cognitive map"
[Tolman, 1948]). It is important to note that no external a priori information
about the external world is given to the system. Learning new locations and

J. Wyatt and J. Demiris (Eds.): EWLR 1999, LNAI 1812, pp. 103–119, 2000.

their links with each other is the core process of the system. In addition, there is also a reflex mechanism for obstacle following. This mechanism takes control of the animat when an obstacle is encountered. The created map is typical for a given environment. It allows the animat to smartly solve various action selection problems in a static world. But the map may be updated so that the same system may work if the environment has changed (door closed or moving person for instance). The other dynamics followed by the system is given by the animat movements in the environment. These movements are either random when the animat is exploring, or given by the planning map when the animat has to reach a goal. Under the neural field formalism (*dynamic approach* to autonomous robotics [Schöner et al., 1995]), we deduce from the planning map a control variable for the animat speed (see [Goetz and Walters, 1997] for a review of control techniques). This enables a smooth approach of the goal.

In the following, we first describe the experimental environment and the planning architecture. Then we show how the planning map may be interpreted in terms of a dynamic approach for speed control. Finally, we present the results about finding a path in a static and then in a changing world.

2   Experiment

The experiments reported in this paper are performed in a simulated environment. However, some of them are already running on a real robot. The animat only sees obstacles and landmarks. It may not see the resource locations. The environment may be as large as 40 animat sizes. Three different energy levels decrease exponentially over time unless re-supplied by a proper source [Gaussier et al., 1998]. Since we are in an animat approach, these levels will be called "food", "water" and "sleep". They are linked with the motivations "hunger", "thirst" and "rest". A more robotic way of putting it in words would be to consider for instance a "charging station", and two different "delivery points", provided the robot gets rewards being at these locations. So, through random exploration, the animat has to discover "food" and "water" places where it may re-supply one of the two levels. From time to time, the animat has to go back to its nest for rest (it knows where it is at start). The animat has to go back to a previously discovered interesting place when it needs to. The animat must therefore solve the dilemma between exploring and reaching a known food and/or water place and/or rest. When on a source location, the animat source level increases by a given amount. When the source is empty, another one appears randomly in the environment. Hence, the animat always has to explore the environment in order to find new potential sources. In addition, we impose that the levels of the essential variables are maintained between two thresholds that define the comfort area. Hence, as soon as the level of one of these variables is below the lower threshold, the animat goes toward a resource place. Conversely, when the level of a particular essential variable is higher than a top threshold, the animat goes away from the resource place. The animat succeeds quite correctly in maintaining the 3 essential variables in the comfort area. A cycle of go and return takes place between the different types of sources. These cycles change only when the animat succeeds, during a random exploration phase, in finding a new resource place, or when a location is depleted. The animat can fail only when new sources appear in areas difficult to explore so that there is a low probability for a single animat to find the solution. In this case, there is a need to have a cooperation between several animats [Gaussier et al., 1997b].

3 Model

We will not fully develop here our navigation model (see [Gaussier et al., 1998], [Gaussier and Zrehen, 1995], [Gaussier et al., 2000] for a complete presentation). What should be reminded is that a location is defined by a set of landmarks and azimuth of each landmark. As the animat explores the environment, it learns new locations when they are different enough from the previous ones. A neuron is coding for each learned position in the environment. The closer the animat to the neuron coding for a position, the higher the response of that neuron. We don't provide any external explicit description of the environment (world model), nor learn what to do for each location in the environment [Burgess et al., 1994]. In order to be able to go back to a learned location, the animat has to keep track of where it was and where it may go from any location. So we need to introduce a kind of map of the environment to be able to perform this task. To do so, we add a group of neurons able to learn the relationships between successively explored places. The temporal proximity being equivalent to a spatial proximity, the system creates a topological representation of its environment. We will call this last group our "cognitive map" (or goal map)(fig. 2 and 1) [Tolman, 1948], [Thinus-Blanc, 1996], [Revel et al., 1998].

Let W_{ij} be the weight associated with the fact that from the place i it is possible to reach directly place j. At the goal level (fig. 1), the motivations activate directly the goal neurons linked with the goal to be achieved (the goal neuron intensity is proportional to the motivation). This activity is then diffused from neuron to neuron through the synaptic weights. Other works also use a diffusion mechanism for planning paths [Connolly et al., 1990], [Bugmann et al., 1995], [Bugmann, 1997]. Our goal diffusion algorithm is the following :

- i_0 *is the goal neuron activated because of a particular motivation. G_{i_0} is its activity.*
- $G_{i_0} \leftarrow 1$ and, $G_i \leftarrow 0 \; \forall i \neq i_0$
- *WHILE the network activity is not stabilized, DO:* $\forall j, G_j \leftarrow \max_i(W_{ij}.G_i)$

where G_i is the value of neuron i (see [Gaussier et al., 1998] for more details on the navigation algorithm). W_{ij} has to be strictly inferior to 1 ($0 < W_{ij} < 1$) to ensure the algorithm convergence. On figure 1 an example of activity diffusion from motivation "A" is presented. The algorithm allows to find the shortest path in the graph. The algorithm is proved to always find the shortest path in

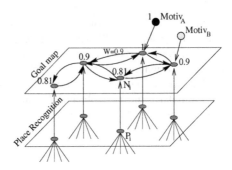

Fig. 1. Global architecture of the planning system. The recognition level allows to identify the situations when the animat arrives in the vicinity of a stored place. These situations are directly linked to the goal level which allows to plan a route from one attractor to the next until the objective. When a motivation "A" activates a goal, a propagation of the information is performed in the direction of all nodes in the graph (here all weights are equal to 0.9).

the graph (it is equivalent to the Bellman-Ford algorithm of graph exploration [Bellman, 1958], [Revel, 1997]).

The animat tries to follow the gradient of neuron activity in this cognitive map to select the next location to reach. The most activated goal or subgoal in the neighborhood of the current animat location is then selected and used to attract the animat in its vicinity. When the animat is close enough to this location, the associated subgoal is inhibited and the animat is attracted by the next subgoal and so on until it reaches the final goal. The principle of this kind of cognitive maps is not new [Arbib and Lieblich, 1977], [Schmajuk and Thieme, 1992],   [Bachelder and Waxman, 1994], [Schölkopf and Mallot, 1994],   [Bugmann et al., 1995],   [Trullier et al., 1997], [Franz et al., 1998]. The novelty in this paper is that our algorithm allows to solve planning problems involving several moving goals in a dynamic environment (the sources may disappear and appear again elsewhere, obstacles may also be moved, landmarks may be hidden). The map is learned and modified on-line and allows to manage contradictory goals. The subgoals correspond to the following situations: the end of an obstacle avoidance (for instance, the animat stores the pathway between two rooms: location of the door) or places badly recognized (activity of all neurons below a defined recognition threshold. The higher this threshold is, the more learned places there are: denser coverage of the open space). We obtain a spatial "paving" of the environment which is almost regular (fig. 2).

Learning of the cognitive map is performed continuously. There is no separation between the learning and the test phases. The links between neurons of the graph are reinforced (hebbian associative learning) for neurons associated with successively recognized places. The learning rule is the following:

$$\frac{dW_{i,j}}{dt} = -\lambda.W_{i,j} + (C + \frac{dR}{dt}).(1 - W_{i,j}).\overline{G_i}.G_j \tag{1}$$

Fig. 2. On the left, reflex behavior of obstacle following for reaching the goal (Initial speed towards the upper left. The animat has a mass giving a momentum (see section 4)). The dots indicate the trajectory followed by the animat. On the right, "cognitive map" built by exploration of the same environment. The landmarks are the crosses on the border. Each circle is a subgoal. The links indicate that the two subgoals have been activated in succession. The subgoals and the learned transitions form the goal or cognitive map.

G_i must be held to a non null value until G_j (with $i \neq j$) is activated by the recognition of the place cell j. This is performed by a time integration of the G_i values represented in the equation by $\overline{G_i}$. $\overline{G_i}$ decreases with time and can be used as a raw measure of the distance between i and j. λ is a very low positive value. It allows forgetting of unused links. $C = 1$ in our simulations. The term $\frac{dR}{dt}$ corresponds to the variation of an external reinforcement signal (negative or positive) that appears when the animat enters or leaves a "difficult" or "dangerous" area.

4   Use of the Planning Map for Speed Control

The planning map is constructed in order to indicate which path to take to reach the goal. In addition, the map may give us another information: how to control the speed of the animat in order to assure to go to the goal and stop on it. Hence the planning map not only indicates where the goal is, but also enables to control how to reach it. For this purpose, we have to consider that the animat has a mass M. At the same time, we also add friction F. To balance this effect, an internal drive is also present. When the animat is exploring, this drive is random in strength and direction. This enables to randomly explore large areas. When it is goal seeking, the animat is going from one subgoal to another until reaching the goal. This is performed by following the gradient of the motivations on each subgoal. So when the animat is going towards the nearest goal (or sub-goal), the direction of the drive is given by the direction of the goal, and the strength may be either independent from the distance to the final goal (hence constant), or depend on this distance (hence depend on the gradient). The equations driving the animat movement are now (we took the same form as in [Schmajuk and Blair, 1992]):

$$v_x(t+1) = v_x(t) - \frac{dt.F.v_x(t)^2}{M} + \frac{dt.drive}{M}$$
$$x(t+1) = x(t) + dt'v_x(t) \tag{2}$$

where the x subscript denotes a projection on the x axis. The same equations hold for the y axis. v is the speed and (x, y) the position of the animat. $F = 5$, $M = 5$ and $drive$ a force (either random or towards the goal). The values of these constants have been chosen in order to exhibit nice "gliding" trajectories illustrating the effect of the external drive in this equation. dt and dt' are small integration constants. In order to stay at the same position (when a goal is reached for instance), the speed must be 0. Hence $drive = 0$ at the goal location. So when the drive is constant, the speed will tend to $\sqrt{\frac{drive}{F}}$ and the animat can not stop on the goal. If the drive depends on the distance to the goal, then one has to make $drive$ tend to 0 as the goal is nearer. In the case where the drive is not constant, it is the difference between the value of the nearest subgoal and the value of the second nearest subgoal which gives the strength of the drive. This value is computed during the diffusion of the motivation from the goal to the subgoals. Using equation 3, this value is $G_n = W^n$, where W is the (fixed) weight between the subgoals and n the distance (expressed in number of subgoals) to the goal. Figure 3 shows the strength of the subgoal G_n versus the distance to the goal (dotted line).

As one can see, as the animat comes close to the goal, the gradient increases. Hence as the animat approaches the goal, the drive towards it increases. This leads to a non null speed when arriving on the goal and an unstable reaching behavior (dotted line in fig. 4). In order to avoid this unstable behavior (or to stop on the goal), the speed must decrease as the animat comes closer to the goal. So the activation value used to compute the gradient is now:

$$H_n = exp - \frac{log(\frac{G_n}{W})^2}{2.\sigma^2} = exp - \frac{log(G_n) - log(W)}{\sigma^2} \tag{3}$$

cp where σ is chosen to be 5.

The speed near the goal is now reduced, enabling a smooth approach (fig. 3). This is particularly suitable if the animat has to stop on the goal.

Comparing the two drive politics, it appears that the first one enables to reach the goal sooner, with a high speed when arriving on the goal. The goal is reached faster because its attraction strength is high, so that the animat does not "glide" as much as in the second case. However, there may be also cases where the goal is not reached at all (see below). In the second case, the animat may take a longer time to reach the goal, but arrives there with a very low speed. However, due to the smaller strength of the attraction it may sometimes "circle around" the goal until it is reached (see fig. 5).

These results are backed by the theoretical framework of dynamical systems. In particular, the way the map is constructed fits very nicely in the "neural field" framework [Amari, 1977], [Schöner et al., 1995]. Indeed, we can extend

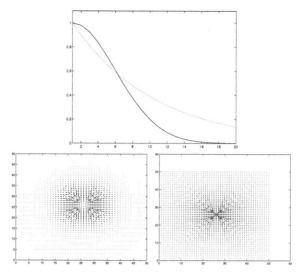

Fig. 3. H_n versus n (solid line), and G_n (dotted line). In the first case, the gradient decreases as the goal is being reached. Whereas in the second case, the gradient increases, leading to a high speed near the goal. The two figures below show left the value of the gradient in the environment when the goal is in the middle of the room. The arrows indicate the direction of the gradient and its strength (left gradient of H_n and right gradient of G_n).

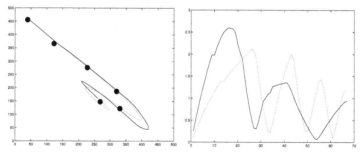

Fig. 4. Left: Trajectory of an animat driven by the gradient computed using H_n (solid line) and G_n (dotted line). The dots are the learned subgoals. The planning map links are not shown. Right: The same conventions are used for displaying the speed versus time in the two same cases. Speed is higher in the second case than in the first one.

the notion of distance expressed in number of subgoals by making it continuous ("behavioral dimension" [Schöner et al., 1995]). So, one can now consider that the goal has an activation field extending over each subgoal. The value of the field at subgoal n is H_n (fig. 3). The same way it is possible to control the heading direction of an animat [Bicho and Schöner, 1997] by controlling its angular velocity, we control the position of the animat through modifications of its speed

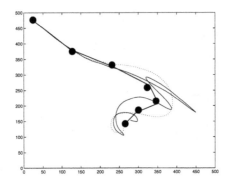

Fig. 5. Trajectory of an animat driven by the gradient computed using H_n (solid line) and G_n (dotted line). The dots are the learned subgoals. The planning map links are shown. The animat reaches the goal sooner in the second case, but with a high speed.

introduced by the gradient of H_n. In this case the goal position is a stable fixed point of the dynamics $\dot{n} = F(n)$ (fig. 6). When $n = 0$, the animat is on the goal (and its speed is 0 either). Moreover, as it is a stable fixed point, we are sure that the animat will converge on it (in asymptotical time theoretically). This is not the case when using a gradient computed with G_n. The animat may not converge on the goal at all because the goal is not a fixed point anymore (fig. 6).

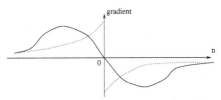

Fig. 6. Gradient of H_n (solid line) versus n and gradient of G_n (dotted line) versus n. In the first case, $n = 0$ is a stable fixed point for the dynamics $\dot{n} = F(n) = \frac{dH_n}{dn}$. In the second case, this not the case.

5   Use of the Planning Map for Finding Paths

We will now show in the next subsections that the animat is able to learn to choose between goals and that the planning map may be used in a changing environment.

5.1   Learning Paths in a Fixed Environment

To allow a measurement of the performances, the environment is chosen in the following experiments as a T-maze. However, this maze is not discretized into squares (other than for statistical results, see below), but allows a continuous computation of the animat positions. The width of a corridor is 7 animat sizes. The need to go to the nest increases twice as fast as the need for food or water. Hence the animat has to go back very often to its nest. In order to suppress the possible biases introduced by an autonomous map building, the cognitive map is learned during a teleoperated exploration of the maze. The operator driving the animat follows the middle of the corridors. The maze and the cognitive map are displayed as the first figure of each experiment shown thereafter. In order to record the preferred paths taken by the animat, we have divided the maze into squares. Each square has the size of an animat. However note that these squares are **not** used in the computation of the movement. They are only used in this statistical analysis. The different figures show histograms of the occupation of each square in the maze. Histograms are computed from the average of 50 different runs for each experiment. One run corresponds to 20000 iterates of the animat behavior in the same T-maze. The animat needs less than 2000 iterates to construct the complete planning map.

Because the activation level of a particular subgoal is the maximum of the back-propagated motivational information, we believed at first that our algorithm was unable to choose correctly between satisfying one motivation or several simultaneous motivations. For instance, we thought it would be unable to always choose the left arm of the T maze fig. 11 that allows to satisfy at the same time 2 motivations (something that the "free flow" architecture of Tyrrell [Tyrrell, 1993] succeeds in doing at the price of local minima problems and a priori simplification in the graph). Of course, it is possible to decide that if a goal is associated with 2 motivations those two motivations are summed before being diffused with our max rule. Unfortunately, if the two interesting sources are not exactly at the same place but are in neighbor areas, this trick cannot be used. If we change the max rule into a simple addition rule we can face deadlock situations because the diffused activity can amplify itself in loop situations or when a node receives the diffusion of a lot of other nodes. So we have tried our algorithm as it is and we have verified it was nevertheless able to find the good solution after a while because of on line weight adaptation.

Figure 7 shows a symmetrical maze. What happens in this case is that the left or the right trajectory is randomly reinforced. Hence the animat goes either to the right arm or to the left arm to drink. But if during a random exploration, it goes in the other arm, it reinforces this path, and therefore enables its use again. Hence, the animat uses only one path until it goes in the other arm and then may use only the other path, and so on.

Figure 8 and 9 show what happens when one of the arms is extended. As expected, the only water source used after a while is the one in the right arm. The exploration of the left arm does not modify enough the weights between the subgoals so as to make the animat choose this water source for drinking. Instead, as the animat goes more often near the water source in the right arm, these links are reinforced.

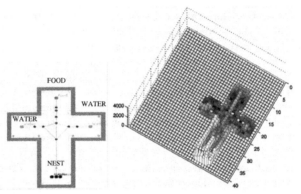

Fig. 7. The figure on the left displays the initial maze, with the nest on the bottom, water sources in the left and the right arm and a food source at the top. In the figure on the right, the height of each square shows the number of time it has been occupied by the animat (dark (resp. light) color indicates low (resp. high) occupation. The number of iterations is 20000. The values have been computed adding 50 different runs. Here, as the maze is symmetrical, no arm is preferred for going to drink.

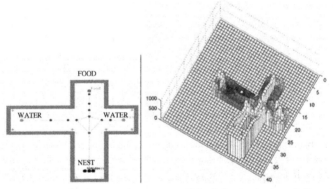

Fig. 8. Same setup as in the previous experiment. The left arm is extended, so that the left water source is farther away from the nest and from the right water source than in Figure 7. Hence, statistically, the animat prefers to go drinking in the right water source.

In the last two experiments (fig. 10 and 11), the arm where the food source is, is shifted to the left. The preferred water source is the one near the food source. When the animat goes eating, it may afterwards explore the maze. Since it is in the left arm, the animat is more likely to go in the left end. Hence the links between the food source and the left water source are reinforced. Only from time to time when the animat has enough time to go to the right end are the links to the right water source reinforced.

The difference between the experiments shown in figures 8, 9, 10 and 11 is the shifting to the left of the upper arm containing the food. This move enables

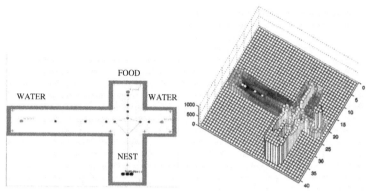

Fig. 9. Same setup as in the previous experiment. The difference with figure 8 is the farther location of the water source in the left arm in this case. As this source is moved away to the left, it gets less used by the animat for drinking.

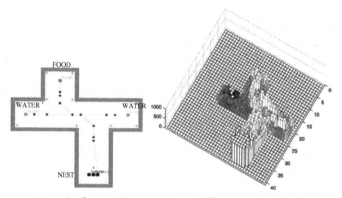

Fig. 10. Same setup as in the previous experiment. The animat visits the right source more often than the left one.

the animat to use the left arm water source more often. And in the extreme (last) case, it is now the left water source which is a lot more often used than the right one. This shows the importance of the reinforcement of used links (and conversely the decrease of the unused links). This property allows the animat to "behave smartly" by not wasting its energy going to far away sources.

5.2 Learning New Paths in a Changing Environment: Preliminary Results

A real environment does not always stay the same. New obstacles may appear or disappear (somebody walking or a door opened or closed, for instance). The change in the environment may first affect the recognition of a place. But this is not the case in our system. Indeed, it stands up the loss (or the addition) of landmarks provided at least half of the landmarks used for learning a location

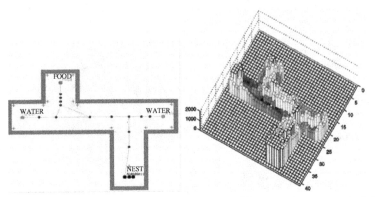

Fig. 11. Same setup as in the previous experiment. The water source on the left, near the food source, is more often used than the water source in the right arm.

are still visible. However, a changing environment may dramatically impair the relevance of a learned cognitive map: some paths may not be used anymore (door closed), some others may be used (door opened). If an obstacle suddenly appears, the reflex mechanism allowing to follow the obstacle enables to nevertheless reach the goal (see figure 2). On the contrary, new paths are found by random exploration of the environment when it is not goal seeking. We have tested our algorithm **as it is** in a changing environment (see figure 12). In this environment, the door opening enabling a direct path may be closed. The animat has learned a cognitive map with the door open. When there are two ways to go from one source to the other one, the direct path is preferred, but the other one is also used (left figure). In the second experiment, the direct path is closed by a door. The animat now goes through the other path.

What makes it possible to change the environment is that the map is constantly updated either by creation of new subgoals (or goals) or by increase (resp. decrease) of the value of used (resp. unused) links. We have shown in the previous section how the reinforcement of some links may lead to the emergence of preferred paths. Now, it is the decrease of the link values which will help "forget" unusable paths. For instance, if the animat has learned a path between two walls and this aperture is suddenly closed (by a door for instance), the cognitive map is not valid anymore. What happens now is a decrease to 0 of the link going "through the wall". The principle for the degeneration of a link is that it is not reinforced anymore because its two ends are not activated in succession anymore. So, the link decreases to 0 due to the passive decay term in equation 1. The 3 typical cases one may encounter when a path may not be used anymore are summarized in figure 13. In the first two cases, one of the ends of the link leading through the wall may not be activated at all. Hence this link degenerates. The only problematic case is the last one presented. Indeed, there, the place cell in the wall may fire either when the animat is on the right or from the left side. Hence the left link, as well as the right link may be reinforced. The solution is

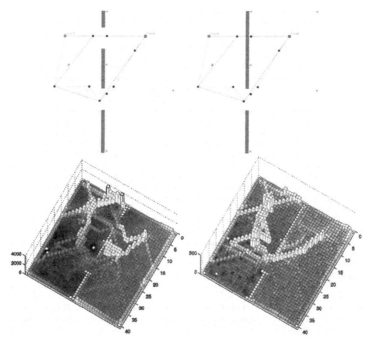

Fig. 12. Same setup as in the previous experiment. In the first experiment (left 2 figures), the animat may go from one source to the other through either door. In the second experiment, the top door is closed. The animat is able to change its behavior and goes through the last open door.

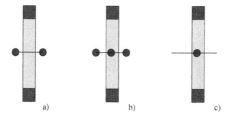

Fig. 13. A door (light hatching) has closed the pathway between two walls. There are 3 different cases depending on where the subgoals were learned when the door was open. In case a), there is no subgoal "in" the door. Since the 2 subgoals on either side of the door can not be activated successively, the link between them degenerates. In case b), for the same reason the links between the subgoals in the middle and the ones outside may not be reinforced. In case c), however the links may be reinforced.

the creation of a new subgoal where the animat hits the door. The situation will then be the same as case b). The creation of this new subgoal (in planning mode specifically, in exploration mode there is no need to create a new subgoal when hitting an obstacle) has not been implemented in the system for the moment.

It is worthwhile to note that after the closure of the door the environment has changed: the landmarks of the other room are not visible anymore. Hence the situation which has led to the creation of a subgoal near the door opening is not valid anymore either. In some cases, these subgoals may not be activated at all (even if the animat is exactly on them) because other subgoals, which correspond to the new environment, have been learned. Indeed, changing the environment also changes the set of locations for which a place cell responses. New place cells are learned corresponding to the new landmark configuration. Their attraction basin extends over the previously created ones. Hence, in the extreme case, a place cell learned in the old environment may not fire at all in the new one, and so no new links may be created with it. However, the old ones still exist (even if their value is close to 0), but may not be reinforced anymore. If the old environment is presented again, these links may be reactivated. The animat must first discover that the path is usable again, and use it for some time before the links values reach a high level again. Now, the same degeneration process happens for the links between place cells created in the second environment. Hence several different "layers" of cognitive maps can appear in the same physical N.N. structure. They may be linked together through some place cells valid in more than one environment, and may be activated when the environment they are coding is presented again. The animat has to try some time before building a new efficient planning map. So, even if it may be a good solution to use a very low passive decay (λ parameter) to store several different maps (memory effect), it also slowers down the process of finding new pathways, when one may not be used anymore. Indeed, the hebbian learning rule we have chosen needs some time before significantly changing the weights. Hence, in order to react faster to a change in the environment, it would be necessary to introduce an active decay mechanism decreasing unused links.

6   Discussion

The navigation and planning system we have presented is able to solve complex action selection problems. However, for the moment, it has to really perform the movements in order to reinforce particular paths. An improvement would be the possibility to internally replay the trajectory used to reach a goal.

The main drawback of the algorithm is the computation of the gradient of G_n. Indeed, in very large environments (with a great number of subgoals) the gradient may be very small. Hence, first, the drive on the speed would be very small too (leading to a long time before reaching the goal). Second, if there is a small noise on the gradient, it would be now impossible to follow it. So there is a need to combine several maps of different joint environments. This means we have to define a planning structure, or plans of plans, in order to address large scale environments. This need is also highlighted by the use of the same neural structure for storing all maps linked with different environments. There will soon be an explosion of the number of neurons needed to code all locations, and moreover a mix up between maps coding for totally different environments. So there is a need to "transfer" the map in another structure where it may be memorized.

Concerning the interpretation of the planning map in terms of activation field, it is important to stress that this process is a top down one. Indeed, the map has first been build using the place recognition mechanism. Now, in return the use of the map gives information for the low level control of the animat. Moreover, this activation field approach could be generalized using the neural field paradigm [Amari, 1977]. Each goal can have a previous defined activation. The activations may be combined using the neural field equations. The advantage is that the animat will stay focused on a goal until either another goal becomes stronger or an external events occurs. This is also very interesting for us because it is easy to introduce in our neural architecture. Following a moving object is already coded that way [Gaussier et al., 1999].

This work was supported by two French GIS contracts on Cognitive Sciences entitled "comparison of control architectures for the problem of action selection" in collaboration with the Animat lab (J.Y. Donnart, A. Guillot, J.A. Meyer), LISC (G. Deffuant) and RFIA (F. Alexandre, H. Frezza), and "Mobile robots as simulation tools for the dynamics of neurobiological models: adaptation and learning for visual navigation tasks" in collaboration with the CRNC (G. Schöner) and the LAAS (R. Chatila).

References

Amari, 1977. Amari, S. (1977). Dynamics of pattern formation in lateral-inhibition type neural fields. *Biological Cybernetics*, 27:77–87.

Arbib and Lieblich, 1977. Arbib, M. and Lieblich, I. (1977). Motivational learning of spatial behavior. In Metzler, J., editor, *Systems Neuroscience*, pages 221–239. Academic Press.

Bachelder and Waxman, 1994. Bachelder, I. and Waxman, A. (1994). Mobile robot visual mapping and localization: A view-based neurocomputational architecture that emulates hippocampal place learning. *Neural Networks*, 7(6/7):1083–1099.

Baloch and Waxman, 1991. Baloch, A. and Waxman, A. (1991). Visual learning, adaptive expectations and behavioral conditionning of the mobile robot mavin. *Neural Networks*, 4:271–302.

Bellman, 1958. Bellman, R. (1958). On a routing problem. *Quarterly of Applied Mathematics*, 16:87–90.

Bicho and Schöner, 1997. Bicho, E. and Schöner, G. (1997). The dynamics approach to autonomous robotics demonstrated on a low-level vehicle platform. *Robotics and Autonomous System*, 21:23–35.

Bugmann, 1997. Bugmann, G. (1997). *Basic Concepts in Neural Networks: A survey*, chapter Chap 5: A Connectionist Approach to Spatial Memory and Planning. Perspectives in Neural Networks. Springer.

Bugmann et al., 1995. Bugmann, G., Taylor, J., and Denham, M. (1995). Route finding by neural nets. In Taylor, J., editor, *Neural Networks*, pages 217–230, Henley-on-Thames. Alfred Waller Ltd.

Burgess et al., 1994. Burgess, N., Recce, M., and O'Keefe, J. (1994). A model of hippocampal function. *Neural Networks*, 7(6/7):1065–1081.

Connolly et al., 1990. Connolly, C., Burns, J., and Weiss, R. (1990). Path planning using laplace's equation. In *International Conference on Robotics and Automation*, pages 2102–2106.

Franz et al., 1998. Franz, M., Schölkopf, B., Mallot, H., and Bülthoff, H. (1998). Learning view graphs for robot navigation. *Autonomous Robots*, 5:111–125.

Gaussier et al., 2000. Gaussier, P., Joulain, C., Banquet, J., Leprêtre, S., and Revel, A. (2000). The visual homing problem: an example of robotics/biology cross fertilization. *Robotics and Autonomous Systems*, 30:155–180.

Gaussier et al., 1997a. Gaussier, P., Joulain, C., Zrehen, S., Banquet, J., and Revel, A. (1997a). Visual navigation in an open environment without map. In *International Conference on Intelligent Robots and Systems - IROS'97*, pages 545–550, Grenoble, France. IEEE/RSJ.

Gaussier et al., 1998. Gaussier, P., Leprêtre, S., Joulain, C., Revel, A., Quoy, M., and Banquet, J. (1998). Animal and robot learning: experiments and models about visual navigation. In *Seventh European Workshop on Learning Robots- EWLR'98*, Edinburgh, UK.

Gaussier et al., 1997b. Gaussier, P., Moga, S., Banquet, J., and Quoy, M. (1997b). From perception-action loops to imitation processes: A bottom-up approach of learning by imitation. In *Socially Intelligent Agents, AAAI fall symposium*, pages 49–54, Boston.

Gaussier et al., 1999. Gaussier, P., Moga, S., Banquet, J., and Quoy, M. (1999). From perception-action loops to imitation processes. *Applied Artificial Intelligence*, 1(7).

Gaussier and Zrehen, 1994. Gaussier, P. and Zrehen, S. (1994). A topological map for on-line learning : Emergence of obstacle avoidance in a mobile robot. In *From Animals to Animats: SAB'94*, pages 282–290, Brighton. MIT Press.

Gaussier and Zrehen, 1995. Gaussier, P. and Zrehen, S. (1995). Perac: A neural architecture to control artificial animals. *Robotics and Autonomous Systems*, 16(2-4):291–320.

Goetz and Walters, 1997. Goetz, P. and Walters, D. (1997). The dynamics of recurrent behavior networks. *Adaptive Behavior*, 6(2):247–283.

Meyer and Wilson, 1991. Meyer, J. and Wilson, S. (1991). From animals to animats. In Press, M., editor, *First International Conference on Simulation of Adaptive Behavior*. Bardford Books.

Millan and Torras, 1992. Millan, J. R. and Torras, C. (1992). A reinforcement connectionist approache to robot path finding in non-maze-like environments. *Machine Learning*, 8:363–395.

Revel, 1997. Revel, A. (1997). *Contrôle d'un robot mobile autonome par une approche neuromimétique*. PhD thesis, U. de Cergy-Pontoise.

Revel et al., 1998. Revel, A., Gaussier, P., Leprêtre, S., and Banquet, J. (1998). Planification versus sensory-motor conditioning: what are the issues ? In Pfeifer, R., Blumberg, B., Meyer, J., and Wilson, S., editors, *From Animals to Animats : Simulation of Adaptive Behavior SAB'98*, pages 129–138. MIT Press.

Schmajuk and Blair, 1992. Schmajuk, N. and Blair, H. (1992). Place learning and the dynamics of spatial navigation: a neural network approach. *Adaptive Behavior*, 1:353–385.

Schmajuk and Thieme, 1992. Schmajuk, N. and Thieme, A. (1992). Purposive behavior and cognitive mapping: a neural network model. *Biological Cybernetics*, 67:165–174.

Schölkopf and Mallot, 1994. Schölkopf, B. and Mallot, H. (1994). View-based cognitive mapping and path-finding. Arbeitsgruppe Bülthoff 7, Max-Planck-Institut für biologische kybernetik.

Schöner et al., 1995. Schöner, G., Dose, M., and Engels, C. (1995). Dynamics of behavior: theory and applications for autonomous robot architectures. *Robotics and Autonomous System*, 16(2-4):213–245.

Thinus-Blanc, 1996. Thinus-Blanc, C. (1996). *Animal Spatial Navigation*. World Scientific.

Tolman, 1948. Tolman, E. (1948). Cognitive maps in rats and men. *The Psychological Review*, 55(4).

Trullier et al., 1997. Trullier, O., Wiener, S., Berthoz, A., and Meyer, J. (1997). Biologically based artificial navigation systems: review and prospects. *Progress in Neurobiology*, 51:483–544.

Tyrrell, 1993. Tyrrell, T. (1993). *Computational Mechanisms for Action Selection*. PhD thesis, University of Edinburgh.

Verschure and Pfeifer, 1992. Verschure, P. and Pfeifer, R. (1992). Categorization, representation, and the dynamics of system-environment interaction. In *From Animals to Animats: SAB'92*, pages 210–217.

Probabilistic and Count Methods in Map Building for Autonomous Mobile Robots

Miguel Rodríguez[1], José Correa[1], Roberto Iglesias[1],
Carlos V. Regueiro[2], and Senén Barro[1]

[1] Dept. Electrónica e Computación,
Universidade de Santiago de Compostela, 15706 Spain,
mrodri@lugo.usc.es, {jcorrea, rober, senen}@dec.usc.es
[2] Dept. Electrónica e Sistemas,
Universidade da Coruña, 15071 Spain,
cvazquez@udc.es

Abstract. In this paper two computationally efficient methods for building a map of the occupancy of a space based on measurements from a ring of ultrasonic sensors are presented. The first is a method based on building a histogram of the occurrence of free and occupied space. The second is based on the calculation of the rate between occupied space measurements with respect to the total. The resulting occupancy maps have been compared with those obtained with other well-known methods, both count as well as Bayes-rule-based ones, in static environments. Free space, occupied space and unknown labels were also compared subsequent to the application of a simple segmentation algorithm. The results obtained gave rise to statistically significant differences between all the different types on comparing the resulting maps. In the case of comparing occupancy labels, no differences were found between the following pairs of methods: RATE and SUM ($p - value = 0.157$), ELFES and RATE ($p - value = 0.600$) and ELFES and SUM ($p - value = 0.593$).

1   Introduction

Two of the basic aims of sensorial perception in navigation tasks are the detection of free space and the construction of maps of the environment. For these two tasks ultrasound (US) sensors have been widely used, due to their cheapness, wide range and low computational cost. Nevertheless, the obtention of a sensor-based environment map is a complicated problem due to the scarce and relatively imprecise information that is obtained from it.

One of the most common approaches to map building using US sensors, and one that has been implemented in real, operative robots, is the setting up of an occupancy grid. Its main advantages are simplicity, low computational cost and human readability. Some of the most widely used approaches to the theme are probabilistic techniques based on Bayes theorem [Elfes, 1989a] [Elfes, 1989b] [Konolige, 1996] [Thrun, 1998], although there are other techniques that are based on fuzzy logic [Gambino, Oriolo and Ulivi, 1996] [Oriolo, Ulivi and

J. Wyatt and J. Demiris (Eds.): EWLR 1999, LNAI 1812, pp. 120–137, 2000.

Fig. 1. NOMAD 200

Vendittelli, 1998], the theory of evidence [Pagac, Nebot and Durrant-Whyte, 1998] [Duckett and Nehmzow, 1998] and count methods [Borenstein and Koren, 1991]. Probabilistic tools revolve around 2 fundamental aspects:

a) Assignation of an occupancy probability to each cell based on a single measurement.

b) Probability accumulation across readings

Nevertheless, in the probability calculation, terms that are difficult to evaluate appear, which are directly dependent on the environment, and which have to be assigned in an arbitrary manner. On the other hand, the map updating process can become computationally expensive. Another common criticism of these techniques is based on the fact that the number assigned by the occupancy probability has neither sufficient quality nor quantity of information on the occupation of a cell.

Based on these considerations, we have developed two algorithms that are very efficient from a computational perspective, and that are based on count methods, where we deal with the number of times that a free or occupied space has been detected in stead of the probability of occupancy. The count idea is not a new one; for example in [Borenstein and Koren, 1991] it was used for the real-time construction of an obstacle map, which later uses an algorithm to avoid obstacles. In spite of not being suitable for our free-space detection needs, a variant of this method adapted to our purpose was used.

All the previously mentioned map construction techniques have been implemented on a NOMAD 200 robot (Fig. 1) and an experiment has been carried out in order to compare the maps obtained under the different processes.

The remainder of this paper is structured in the following manner: firstly, the different map construction methods are presented; then the experimental design used for the comparison of the different techniques are explained, and the results obtained are given, and finally, the results obtained are discussed.

1.1 Related Work

The need for an environment representation has given rise to several designs of map building methods for navigation purposes, starting with those that use the environment itself as its own representation, hence only the current sensor readings are considered in order to drive the current actuator commands, as in Brooks' subsumption architecture [Brooks, 1986].

Other authors use the sensor-input history to build an internal environment map, which is used by navigation algorithms to generate the motor outputs, thus the current behaviour of the robot is driven not only by the current sensor input, but by all the inputs in the input history. In this context, Gasós and Martín [Gasós and Martín, 1997] generated maps that consisted of a fuzzy geometric primitives database that was updated with the new sensor US readings. Kuypers and Byun [Kuipers and Byun, 1991] used the US sensor readings to generate a hierarchical map with several description levels (control level, geometric level and topological level) in which the central part is the topological network description. The arcs in the topological map are assigned to a control behaviour and geometric information can be added to all the components of the topological graph. Nehmzow and Smithers [Nehmzow and Smithers, 1991] used self-organized networks to distinguish space locations. The input vectors to the network were obtained from both sensor measurements and motor commands. Koenig and Simmons [Koenig and Simmons, 1998] used Markov models to represent the perception process. Yamauchi and Beer [Yamauchi and Beer, 1996] used a method that built a so-called Adaptive Place Network, in which a graph-based topological map is combined with metric information. In this context, occupancy grids are used to recalibrate dead reckoning.

Aside from probabilistic and histogramic methods, other occupancy grid map building techniques were also implemented, including those based on fuzzy logic [Gambino, Oriolo and Ulivi, 1996] [Oriolo, Ulivi and Vendittelli, 1998] and the theory of evidence [Pagac, Nebot and Durrant-Whyte, 1998] [Duckett and Nehmzow, 1998].

2   Bayesian Models

There are two fundamental aspects in probabilistic occupancy grid construction methods [Elfes, 1989a] [Konolige, 1996] [Thrun, 1998]: the sensor model, i.e. the assignation of the occupancy probabilities of each cell within the range of a sensor for a single sensorial measurement, and the accumulation of these individual probabilities from multiple measurements.

The most common probability assignation procedure takes the form of a Gaussian sensor model, in which terms that modify the Gauss distribution according to the distance and angular position with respect to the axis of the sonar beam appear. Thus a typical expression would be:

$$p(D \mid r, \theta) = \frac{\alpha(r)}{2\pi\delta(r)\sigma} e^{-\frac{1}{2}(\frac{r-D}{\delta})^2 - \frac{1}{2}\frac{\theta^2}{\sigma^2}} \tag{1}$$

where $p(D \mid r, \theta)$ is the probability density function for the event "**obtain a measurement D the objective being situated at a distance r and forming an angle θ with the axis of the sonar beam**". Here, $\alpha(r)$ and $\delta(r)$ represent factors that modulate normal distribution with distance, and σ represents the typical deviation for the angular deviation (values in the region of $11°$) [Konolige, 1996]. Other authors [Thrun, 1998] use neural networks for the initial assignation of probabilities.

Another characteristic of these methods is the manner of representing uncertainty on an arc of the sector in which the echo is generated. The introduction of the factor $e^{-\frac{1}{2}\frac{\theta^2}{\sigma^2}}$ in (1) or of the factor $(1 - \frac{\theta^2}{\beta^2})$ in other cases, aims to correct this uncertainty by assigning higher probability values to the central parts of the arc, as opposed to the lateral ones. Nevertheless, the introduction of these factors is heuristic, and there are a multitude of situations in which this uncertainty cannot be modelled by means of this procedure.

The updating of the probability assigned to a cell is carried out by means of the Bayes rule (2)

$$P(A \mid B) = \frac{P(B \mid A)P(A)}{P(B)} \tag{2}$$

where A and B are two events, $P(A)$ is the probability of A, $P(B)$ is the probability B, $P(B \mid A)$ represent the probability of B conditioned to A and $P(A \mid B)$ represent the probability of A conditioned to B. If we represent A as the event $C = \{$the cell C is occupied$\}$, and B as the event $[X = D] = \{$an echo has been obtained at a distance D$\}$, then (2) will lead to

$$P(C \mid [X{=}D]) = \frac{p([X{=}D] \mid C)P(C)}{p([X{=}D] \mid C)P(C) + p([X{=}D] \mid \overline{C})P(\overline{C})} \tag{3}$$

where the term $p([X = D] \mid C)$ would be represented by (1), $P(C)$ would represent the probability prior to the current reading, and $P(C \mid [X = D])$ is the posterior occupancy probability.

In other references [Konolige, 1996], the Gaussian curve (1) is substituted by curves taking the form $y(\theta, r) = (1 - \frac{r^2}{R^2})(1 - \frac{\theta^2}{\beta^2})$, but the Bayesian basis is maintained. One difficulty with regard to these methods lies in finding a suitable distribution for the term $p([X = D] \mid \overline{C})$. This represents the probability of obtaining a reading of occupancy when the cell is empty. An exact calculus involves all the possible states of all the cells in the map, so computations over the 2^N possible environment states, where N is the number of cells in the map, are necessary.

2.1   ELFES Method

Two Bayesian methods found in the literature were implemented: ELFES [Elfes, 1989a] [Elfes, 1989b] and MURIEL [Konolige, 1996], both of them particular cases of the general Bayesian technique described above, with slight differences in sensor models (1) and probability accumulation (3).

In ELFES [Elfes, 1989a] [Elfes, 1989b] the terms $\alpha(r)$ and $\delta(r)$ in (1) are constant and heuristically set to 1 and 0.02, respectively, which means that no distance dependency is assumed. In the probability accumulation procedure it is assumed that $P([X = D] \mid \overline{C}) = 1 - P([X = D] \mid C)$, which means that the probability of obtaining a value of occupancy at a distance D, the cell being empty, is assumed to be equal to the probability of not obtaining a value of occupancy at a distance D, the cell being busy.

Probability values below 0.1 were set to 0.1 and probability values above 0.9 to 0.9, with the aim of avoiding not being able to change non-occupancy and occupancy certainties (0 and 1) that are inherent in the Bayes rule (if the probability of the occupancy of a cell is 1 or 0, then this value will be permanent as a consequence of the expression (2)). This allow this method to respond to changes in the environment, or to fix sensor misreadings.

To ilustrate how this method works, we will consider the occupancy values of a given cell before and after the sensor reading, and examine two particular cases:

i) The cell is between the echo and the robot, and further away from the echo. In this case the second term in the denominator of (3) dominates and $P(C \mid [X = D])$ takes a value that is closer to 0 than the old one.

ii) The cell is close to the echo. In this case $p([X = D] \mid C)$ in (3) is higher than $p([X = D] \mid \overline{C})$, and $P(C \mid [X = D])$ will take a value that is closer to 1 than the old one.

2.2   MURIEL Method

MURIEL [Konolige, 1996] is a sophisticated probabilistic method that includes independent sensor data collection, specular reflection filtering and Bayesian probability accumulation. Both independent data collection and specular reflection filtering were developed under the static world hypothesis, and may have unwanted side effects in dynamic environments. As these two features are additional to the probability accumulation, and equivalent mechanisms can be proposed for the rest of the methods, a modified version in which these features were not present was implemented for comparation purposes.

This method starts with a modified Gaussian sensor model in which the possibility of the existence of multiple targets is considered. The expression postulated for the probability density of obtaining a measurement X=D given the cell occupied is

$$p([X = D]|C) = \frac{\alpha(r_i)}{\sqrt{2\pi}\delta(r_i)} e^{-\frac{1}{2}\frac{(D-r_i)^2}{\delta(r_i)^2} - \frac{1}{2}\frac{\theta_i^2}{\sigma^2}} + F \tag{4}$$

being F a small constant representing a constant probability of specular re-flection. In this sensor model of this method (4), distance dependent terms, $\alpha(r)$ and $\delta(r)$, are heuristically introduced as $\alpha(r) = 0.6(1 - \min(1, .25r))$ and $\delta(r) = 0.01 + 0.015r$, and it is assumed that the term $p([X = D] \mid \overline{C})$ in the probability accumulation procedure is the heuristic quantity F.

Given the existence of multiple targets, and given the possibility of obtaining several echoes from one single sensor reading, the event that the detected echo is the nearest one is differentiated. The following notation is used

$[r \rhd D]$: No return less than D

$[r @ D]$: $[r = D]$ and $[r \rhd D]$

Under the hypothesis of independence between $[r = D]$ and $[r \rhd D]$ we obtain the following expression for the posterior likelihood

$$\lambda([r@D|C]) = \frac{P(r@D|C)}{P(r@D|\overline{C})} = \frac{p(r = D|C)P(r \rhd D|C)}{p(r = D|\overline{C})P(r \rhd D|\overline{C})} \tag{5}$$

where $P(r \rhd D|Q) = 1 - \int_0^D \int_0^{2\pi} p(r = x|Q) dx d\theta$, with $Q = C$ or $\overline{C}$. An on-axis plot of the log likelihood is shown in Fig. 2.

These individual likelihoods, measured for each cell, are assumed to be inde-pendent across readings, so the posterior odds for occupancy of a given cell

$$O(C|[x = D]) = \frac{p(C|[X = D])}{p(\overline{C}|[X = D])} = \frac{p([X@D]|C)}{p(|[X@D]|\overline{C})} \frac{P(C)}{P(\overline{C})} = \lambda([X@D]|C)O(C) \tag{6}$$

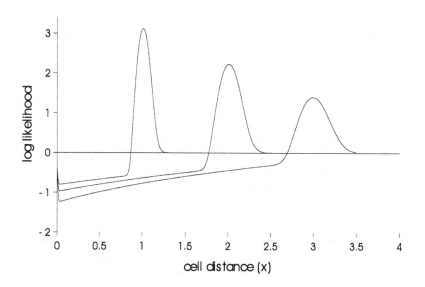

Fig. 2. On axis log likelihood for targets at 1, 2 and 3 m.

hence the log likelihood is accumulated using the recursive equation

$$log(O(C|[x = D])) = log(\lambda([X = D]|C)) + log(O(C)) \tag{7}$$

Qualitatively, this works as follows: Let us suppose that we obtain a sensor reading X=D, and let us consider the posterior odds of a given cell. The posterior odds will be given by (7), and will be the addition of the prior odds and the reading log likelihood (Fig. 2). If the cell is located between the robot and the target, the log likelihood is then less than 0, so the posterior odds will decrease. If the cell is close to the target, the log likelihood will be higher than 1 and the posterior odds will increase. If the cell is after the target, the log likelihood will be close to 0 and, so the posterior odds will remain aproximately constant.

3   Count Based Models

In this category we include those methods that assign a degree of occupancy to each cell in the map, based on the number of times that it is detected as being occupied, and the number of times that it is detected as being empty. In the same manner as in the case of the Bayesian models, these methods comprise a sensor model and an accumulation method.

3.1   SUM and RATE Methods

In our methods we have constructed a very simple sensor model. Each time that a sonar echo is perceived we assign the value $v_t[i, j] = -1$ to the cells c_{ij} situated between the transmitter and the possible obstacle, and the value $v_t[i, j] = +1$ to the cells situated on the boundary sector of the sonar cone, independently of its position on it (Fig. 3). Cells above 3 meters from the centre of the robot are not updated.

Possible speculative reflections were not considered, nor the distance of the ultrasonic source to the possible objective. We have tested sensor models that did deal with this last aspect, but these did not significantly improve the approach that we present here. Rewarding the centre of the segment to the detriment of the outermost points has also been tried (with the function $1 - \frac{\theta}{\beta}$), but no apparent improvement in the maps obtained was observed.

In the SUM model the occupancy value of each cell $C[i, j]$ of the map is obtained by algebraically adding the different observations

$$C_{t+1}[i, j] = C_t[i, j] + v_t[i, j] \tag{8}$$

truncating to a maximum cell occupancy value of 127 and a minimum one of -128. The initial value for all cells is 0, i.e., maximum uncertainty.

This method has a very low computational cost, with a great deal of inertia in exchange. That is, if a cell has been shown to be "free of occupancy" over a series of n observations, at least n new observations in which it is shown to be "occupied" are needed for it to lose the character of "free of occupancy" and for

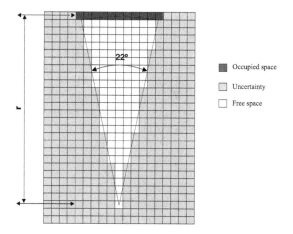

Fig. 3. Sensor model used in SUM and RATE methods, showing the distribution of free and occupied space with regard to one ultrasound measurement.

it to be shown with the characteristics of maximum uncertainty, and a further n observations, in which it must carry on appearing as being occupied, are needed in order for it to classed as "occupied with the degree of certainty n". This is why the method is very efficient in determining static characteristics of maps, by filtering moving obstacles and errors. In any case, the maximum and minimum occupancy values are a design specification, and it will always be possible to adjust the inertia of the map.

In the RATE method we maintain two matrices of values denominated $Occ[i, j]$ and $Visits[i, j]$, so that each time that the sensor model returns a value $v_t[i, j] = -1$ we will increase $Visits_t[i, j]$ and each time that $v_t[i, j] = 1$ is obtained we simultaneously increase $Occ_t[i, j]$ and $Visits_t[i, j]$. The occupancy value for each cell is obtained as the relative frequency

$$C_t[i, j] = \frac{Occ_t[i, j]}{Visits_t[i, j]} \tag{9}$$

which can be considered as a probability.

From a computational perspective, this method has a higher cost than the previous one, as it has to maintain two matrices per map in place of one. In exchange, compared with probabilistic methods, it has the advantage of not involving calculi with transcendent functions.

3.2   HISTOGRAM Method

This method is an adaptation of a technique proposed by Borenstein and Koren [Borenstein and Koren, 1991] designed to detect obstacles online. The original technique increments one single cell in the grid by three units for each range reading. The incremented cell is the one that corresponds to the measured distance r and lies in the acoustic axis of the sensor (black cell in Fig. 4). The cells lying in the acoustic axis with distances less than the measured value are decremented one unit (white cells in Fig. 4). The maximum value accumulated in a cell is 15 and the minimum value is 0, so this method cannot distinguish between free and unexplored space. We have implemented a modified method that maintains the original sensor model, but uses (8) as an accumulation procedure in order to provide free-space information.

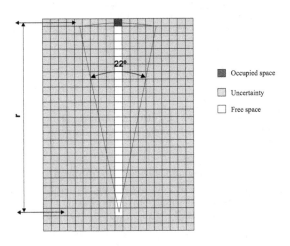

Fig. 4. Sensor model used in HISTOGRAM method [Borenstein and Koren, 1991], showing the distribution of free and occupied space with regard to one ultrasound measurement.

4   Experimental Results

In order to construct the maps that are presented in this paper we have used the following methodology; 1800 data were collected from an excursion of the NO-MAD 200 robot in an indoor environment. During the excursion the robot was autonomously moved by a wall following agent (see [Iglesias, Regueiro, Correa and Barro, 1997] [Iglesias, Regueiro, Correa and Barro, 1998] for details) which followed the wall on the right, and travelled the length of the corridor only once. Each datum consisted of 16 distance measurements from an ultrasonic sensor ring, and their dead reckoning data associated, stored at a frequency of 8 Hz.

The route started from a laboratory, crossed a 1.6 m. wide door and then followed a 2 m. wide corridor to the right, crossed a 2.7 m. × 5.2 m. hall, followed straight on for approximately 3 m. and then stopped. The centre of the laboratory was free, while there was several obstacles close to the walls.

The corridor had three doors on the left, close to the hall, and the hall had an obstacle (coffee machine) on the right, and a corridor and a door on the left. A detail of the route can be found in Fig. 5. The average speed was 0.1 m/s and the total distance covered was about 30 m.

The different methods are applied to this set of data in order to obtain the occupancy values of the 180 × 230 square cells, 0.10 m. wide, in a regular grid. In order to visualize the maps, the occupancy values have been linearly transformed to a grey scale in the range of $[0, 255]$. The dimensions of several features in the environment were measured in each map and compared with the actual values measured directly in the environment. The distances tested were the width and length of the corridor and the width and length of the coffee machine. All distances were obtained manually several times, and at different places, and mean value and standard deviation calculated. Table 1 shows the results of this comparation.

Fig. 5. View of the corridor travelled along by the robot in the experiment.

Figures 6 and 7 show the maps obtained with the two Bayesian methods implemented. As can be seen the definition of the walls is poor.

Figure 8 shows the occupancy map obtained with the SUM method. The cleanness and definition are noteworthy, both in the free space as well as obstacles. The doors along the route can also be seen.

The map obtained with the RATE method is shown in Fig. 9. This map is of a poorer quality than the previous one. Contrary to what happened in the former, the uncertainty error in the sector are very difficult to eliminate and they cast a shadow over the part of the corridor which really corresponds to free space. Comparing the two methods, which have the sensor model in common, the determining factor in order to decide on the occupancy of a cell was a "**democratic vote**" on " **how many in favour, how many against**

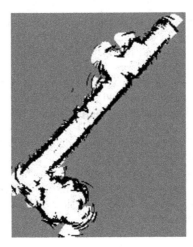

Fig. 6. Map of the reference environment built according to the MURIEL method [Konolige, 1996].

Fig. 7. Map of the reference environment built according to the ELFES method [Elfes, 1989a].

and ... the majority decides" in place of "we have this percentage of the vote in favour of the cell being occupied"

The map in Fig. 10 was obtained by means of the HISTOGRAM method. This method detects obstacles well, and as far as free space is concerned, apparently it signals this more poorly than the previous methods. Nevertheless, we have verified that, in the same manner as the RATE method, this can be

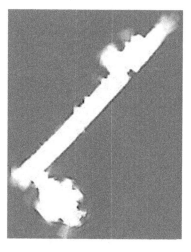

Fig. 8. Occupancy map built according to the SUM method.

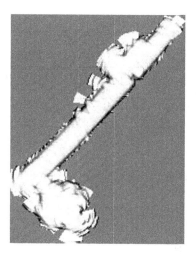

Fig. 9. Occupancy map built according to the RATE method.

avoided by simply considering those values under a certain threshold to be free spaces. None of the methods perform well in the case of the robot remaining idle for long periods of time.

The results of the comparation of several environment dimensions measured in each map versus the actual distances is shown in Table 1, in which it is shown that all the distances measured lie within the measurement error range.

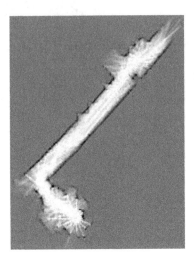

Fig. 10. Occupancy map built according to the HISTOGRAM method (see text for details).

Table 1. Comparation between real environment dimensions and map dimensions.

	Actual dist	SUM	RELATIVE	ELFES	MURIEL	HISTO
corridor width	2.00 m	1.94±0.07	2.01±0.05	1.92±0.08	1.66±0.13	1.97±0.07
corridor length	13.30 m	13.2±0.11	13.2±0.10	13.16±0.07	13.1±0.11	13.27±0.04
obstacule width	0.68 m	0.74±0.11	0.73±0.38	0.75±0.3	0.92±0.24	0.62±0.11
obstacule length	0.58 m	0.72±0.14	0.65±0.07	0.57±0.06	0.79±0.26	0.55±0.11

With the aim of emphasizing obstacle detection in the SUM method we have strengthened the occupancy value establishing it at +3, obtaining the map in Fig. 11.

5   Statistical Analysis

The maps obtained with the five different treatments are, at first view, very different, as can be seen in the figures. These differences were statistically confirmed in the following manner. From each map the value of each cell of the map linearly transformed to the range $[-128, 127]$ are stored. The data from each data set do not follow a known distribution, and for this reason any test to be used for comparing them must be non-parametric. Moreover, all maps were generated from the same set of sensorial data, due to which the statistical test has to take into account that the data set are related.

Bearing this in mind, we decided to apply the non-parametric Friedman test for 5 related treatments [Mendenhall, Scheaffer and Wackerly, 1986] to the five

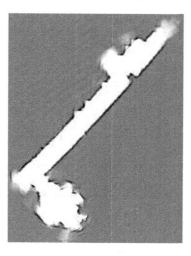

Fig. 11. Occupancy map built according to the SUM method, assigning $v_t[i,j] = +3$, for obstacles.

data sets obtained. In this test the aim is to prove that the five maps are identical (null hypothesis or H_0) as opposed to the hypothesis that at least two of the maps are different (alternative hypothesis or H_a). The result was that there were significant differences ($p - value < 0.001$).

With equality between the set of five maps ruled out, we went on to search for similarities between the ten possible pairs formed on the basis of the five maps. In order to do this the Wilcoxon test for two related samples was applied. Table 2 shows all the $p - values$ obtained together with their statistical verification (Z). As can be seen, the result was also significant ($p - value < 0.001$) in all these cases. Thus, even though we were working with the same set of data obtained by the ultrasonic sensors, the results of the different methods cannot be considered to be similar.

Nevertheless, as a fundamental objective of map construction is the differentiation between free and occupied space, we have proceeded to carry out a second statistical test, aiming to decide with this whether all or some of the methods show this characteristic in the same manner. For this, firstly, the map data were transformed in the following way: each positive value in the new file is assigned the value $+1$ (*occupied*), each negative value is assigned the value -1 (*free*) and the null values (*unknown*) have been maintained.

Statistical analysis for these segmented maps (table 3) throws up significant differences maps in all cases except SUM-RATE ($p - value = 0.157$), ELFES-RATE ($p - value = 0.600$) and ELFES-SUM ($p - value = 0.593$). That is to say that with the objective of labelling the space as free or occupied the given pairs of methods are similar.

Table 2. Verification statistics (Z) and $p-value$ (in brackets) of the Wilcoxon test for two related samples, applied to the maps created with each of the five methods used.

		ELFES	MURIEL	SUM	RATE
MURIEL	-19.031 (0.000)				
SUM	-9.778 (0.000)	-2.759 (0.006)			
RATE	-38.330 (0.000)	-29.788 (0.000)	-9.174 (0.000)		
HISTOGRAM	-61.236 (0.000)	-56.229 (0.000)	-77.317 (0.000)	-63.072 (0.000)	

Table 3. Verification statistics (Z) and $p-value$ (in brackets) of the Wilcoxon test for two related samples, applied to the map data after transformation into free, occupied and unknown spaces.

		ELFES	MURIEL	SUM	RATE
MURIEL	-10.192 (0.000)				
SUM	-0.534 **(0.593)**	-6.756 (0.006)			
RATE	-0.524 **(0.600)**	-6.764 (0.000)	-1.414 **(0.157)**		
HISTOGRAM	-31.520 (0.000)	-26.836 (0.000)	-33.588 (0.000)	-33.607 (0.000)	

6   Discussion

In this work we present two new count based methods for building occupancy maps. One of them, SUM, builds an histogram of the occurrence of free and occupied space. The other, RATE, is based on the calculation of the rate between the number of times one cell is detected as occupied space and the number of times a measurement is got for that cell. The main advantage of count based methods is their very low computational cost, so they can operate in real time on mobile robots where the computational resources are always limited.

In order to test these algorithms, two well known probabilistic methods based on the Bayes rule, ELFES and a variant of Konolige's MURIEL in which independent data collection and specular reflection filtering were not present, were implemented, as well as, one modified version of the histogramic algorithm. Visually comparing all the maps with the real environment results that maps obtained using count methods lead to more realistic maps than probabilistic ones.

Among count methods, HISTOGRAM leads to worse results differentiating free space. The reason is the smaller amount of cells for which an occupancy value is obtained for each sensor reading. The comparation between the distances visually measured in the maps and the actual dimensions shows that in all cases the actual dimension lies within the measurement error range.

Aside from visual comparations, statistical results show that none of the five techniques is equal to any other in occupancy map building. However, if the cells are labelled as occupied, free or unknown, there are no significant differences between SUM, ELFES and RATE methods. This is remarkable, since for navigation tasks the important point is to determine whether or not there is free space for moving the robot. For these tasks, it would be possible to substitute the computationally expensive ELFES method by the simpler and more efficient SUM or RATE.

Count methods have to make strong assumptions in sensor model and accumulation methods, both stated without proof, but the effect of inaccuracies in these assumptions are expected to accumulate incoherently and tend to cancel each other out. In the same way, Bayesian methods are not free of such assumptions, related to the practical difficulties in computing the term $p([X = D] \mid \overline{C})$ in the accumulation mechanism (3), the exact calculation of which would involve computation over all possible environment configurations. The sensor model in this case is also given with no proof.

Nevertheless, all methods for building occupancy grids assume that occupancy values of all measurements are independent for all cells. The rationale is that, as the robot is mobile, occupancy measurements obtained from different positions generate errors that distribute randomly and tend to annul. This hypothesis is, however, frequently violated in practice, and it was observed that some configurations of trajectories and obstacles generate systematic errors that accumulate coherently, mainly, in walls and obstacles detected far away. This effect shows up in different ways depending on the map building method employed, but it is always present. In order to avoid this behavior, it may be necessary to modify the models by considering the dependency between adjacent cells, and measurements taken from adjacent locations.

There is an additional problem related to odometry accumulative errors. These errors lead to erroneous robot position readings in the long term, while they are quite reliable in short displacements. This problem has been addressed by different authors recalibrating the odometry with position estimations based on sensor readings [Thrun, 1998] [Duckett and Nehmzow, 1998]. Occupancy measurements are only possible inside the range covered by the sensors, hence occupancy measurements in a particular cell were taken from locations close to it. In this experiment the trajectory followed was uniform and passed each point only once, so we can assert that all occupancy measurements in a given cell came from positions reached by the robot close in time, and the odometry error is not expected to accumulate excessively. The odometric long term error effect shows up in all maps as a deviation from the straight line in the shape of the walls.

We are currently developing algorithms which use the occupancy grid and measurements from a laser range finder to detect natural landmarks (i.e., doors and corners), which can be used to recalibrate the odometry [Duckett and Nehmzow, 1998] and to recognize regions. The occupancy grid can also be used to find which regions need to be explored or revisited in order to obtain more information about them.

Acknowledgments

This work has been possible thanks to the project XUGA20608B97, and the availability of a Nomad 200 mobile robot acquired through an infrastructure project, both funded by the Xunta de Galicia.

References

[Brooks, 1986]Brooks R.A.(1996) A robust layered controlfor a mobile robot *IEEE Journal of Robotics and Automation*,RA-2(1):14-23.

[Borenstein and Koren, 1991]Borenstein, J. and Koren, Y. (1991) Histogramic in-motion mapping for mobile robot obstacle avoidance. *IEEE Journal of Robotics and Automation*, 7(4):535–539.

[Duckett and Nehmzow, 1998]Duckett, T. and Nehmzow, U. (1998) Mobile robot self-localisation and measurement of performance in middle-scale environments. *Robotics and Autonomous Systems*, 24:57–69.

[Elfes, 1989a]Elfes, A. (1989) Occupancy grids: A Probabilistic Framework for Mobile Robot Perception and Navigation. PhD. thesis, Electrical and Computer Engineering. Dept./Robotics Inst. Carnegie Mellon Univ.

[Elfes, 1989b]Elfes, A. (1989) Using Occupancy Grids for Mobile Robot Perception and Navigation. *IEEE Computer Magazine*, 22(6):46–57.

[Gambino, Oriolo and Ulivi, 1996]Gambino, F., Oriolo, G. and Ulivi, G. (1996) A comparison of three uncertainty calculus techniques for ultrasonic map building. *SPIE Int. Symp. Aerospace/Defense sensing contr.*, Orlando, FL., pages 249–260.

[Gasós and Martín, 1997]Gasós, J. and Martín, A. (1997) Mobile robot localization using fuzzy maps. In Martin, T. and Ralescu, A., editors, *Fuzzy Logic in Artificial Intelligence*, 1188, Springer-Verlag, 207–224.

[Iglesias, Regueiro, Correa and Barro, 1998]Iglesias, R., Regueiro, C.V., Correa, J. and Barro, S. (1998) Supervised Reinforcement Learning: Application to a Wall Following Behaviour in a Mobile Robot, In Pobil, A.P., Mira, J. and Ali, M., editors, *Lecture Notes in Artificial Intelligence*, 1416, Springer-Verlag, pages 300–309.

[Iglesias, Regueiro, Correa and Barro, 1997]Iglesias, R., Regueiro, C.V., Correa, J. and Barro, S. (1997) Implementation of a Basic Reactive Behavior in Mobile Robotics Through Artificial Neural Networks, In Mira, J., Moreno-Diaz, R. and Cabestany, J., editors, *Lecture Notes in Computer Science*, 1240, Springer-Verlag, pages 1364-1373.

[Koenig and Simmons, 1998]Koenig, S. and Simmons, R.G. (1998) Xavier: A Robot Navigation Architecture Based on Partially Observable Markov Decision Process Models In Kortenkamp, D., Bonasso, R.P. and Murphy, R., editors, *Artificial Intelligence and Mobile Robots: Case Studies of Successful Robot Systems.* AAAI Press / The MIT Press.

[Konolige, 1996]Konolige, K. (1996) A Refined Method for Occupancy Grid Interpretation. In Dorst, L., van Lambalgen, M., and Voorbraak, F., editors, *Lecture Nores in Artificial Intelligence*, 1093, Springer-Verlag, pages 338–352.

[Kuipers and Byun, 1991]Kuipers, B.J. and Byun, Y.T., (1991) A robot exploration and mapping strategy based on a semantic hierarchy of spatial representations. *Robotics and Autonomous Systems*, 8:47–63.

[Mendenhall, Scheaffer and Wackerly, 1986]Mendenhall, W., Scheaffer, R.L. and Wackerly, D.D. (1986) *Mathematical Statistics with Applications.* Ed. PWS Publishers.

[Nehmzow and Smithers, 1991]Nehmzow, U. and Smithers, T. (1991) Mapbuilding using Self-Organising Networks in Really Useful Robots In Meyer, J.A. and Wilson, S., editors, *From Animals to Animats, Proc. 1st conference on Simulation on Adaptive Behavior*, MIT Press.

[Oriolo, Ulivi and Vendittelli, 1998]Oriolo, G., Ulivi, G. and Vendittelli, M. (1998) Real-Time Map Building and Navigation for Autonomous Robots in Unknown Environments *IEEE Transactions on Systems, Man, and Cybernetics*, 28(3):316–333.

[Pagac, Nebot and Durrant-Whyte, 1998]Pagac, D., Nebot, M. and Durrant-Whyte, H. (1998) An Evidential Approach to Map-Building for Autonomous Vehicles. *IEEE Transactions on Robotics and Automation*, 14(4).

[Thrun, 1998]Thrun, S. (1998) Learning Metric-Topological Maps for Indoor Mobile Robot Navigation. *AI Journal*, 99(1):21–71.

[Yamauchi and Beer, 1996]Yamauchi, B. and Beer, R. (1996) Spatial Learning for Navigation in Dynamic Environments *IEEE Transactions on Systems, Man, and Cybernetics*, 26(3):496–505.

Biologically-Inspired Visual Landmark Learning for Mobile Robots

Giovanni Bianco[1] and Riccardo Cassinis[2]

[1] University of Verona, Servizio Informatico, Via San Francesco 22,
I-37138 Verona, Italy
bianco@chiostro.univr.it

[2] University of Brescia, Dept. of Elect. for Automation, Via Branze 38,
I-25123 Brescia, Italy
Riccardo.Cassinis@unibs.it

Abstract. This paper presents a biologically-inspired method for selecting visual landmarks which are suitable for navigating within not
pre-engineered environments. A landmark is a region of the goal image
which is chosen according to its reliability measured through a phase
called *Turn Back and Look (TBL)*. This mimics the learning behavior of
some social insects. The TBL phase affects the conservativeness of the
vector field thus allowing us to compute the visual potential function
which drives the navigation to the goal. Furthermore, the conservativeness of the navigation vector field allows us to assess if the learning
phase has produced good landmarks. The presence of a potential function means that classical control theory principles based on the Lyapunov
functions can be applied to assess the robustness of the navigation strategy. Results of experiments using a *Nomad200* mobile robot and a color
camera are presented throughout the paper.

1 Introduction

Animals, including insects, are proficient in navigating and, in general, several
biological ways of solving navigational tasks seem to be promising for robotics
applications. The different methods of navigating have been recently studied and
categorized as [32]: *guidance, place recognition - triggered response, topological*
and *metric* navigations. In order to perform such tasks animals usually deal with
identifiable objects in the environment called *landmarks* [33].

The use of landmarks in robotics has been extensively studied [8,31]. Basically, a landmark needs to possess characteristics such as the *stationarity, reliability* in recognition, and *uniqueness*. These properties must be matched with the
nature of a landmark: landmarks can be *artificial* or *natural*. Of course it is much
easier to deal with artificial landmarks instead of dealing with natural ones, but
the latter are more appealing because their use requires no engineering of the
environment. However, a general method of dealing with natural landmarks still
remains to be introduced. The main problem lies in the selection of the most
suitable landmarks [26,31].

J. Wyatt and J. Demiris (Eds.): EWLR 1999, LNAI 1812, pp. 138–164, 2000.

Recently it has been discovered that wasps and bees perform specific flights during the first journey to a new place to learn color, shape and distance of landmarks. Such flights are termed *Turn Back and Look (TBL)* [22]. Once the place has been recognized using landmarks, insects can then accomplish navigation actions accordingly. The Cartwright and Collett model [9] is one of the main methods of navigating [32].

The aim of this paper is to describe the learning system of a biologically-inspired navigation method based on natural visual landmarks. The introduced system will select natural landmarks from the environment adopting the TBL phase (section 2).

From the selected landmarks suitable navigation movements will be computed (section 3) and in Sect. 4 the guidance principle and how this can be influenced by the TBL phase will be addressed. The measurable effects of TBL (namely, the *conservativeness* of the navigation vector field) is a way to assess the quality of the landmarks chosen by the learning phase.

The conservativeness of the field also permits to compute the (unique) potential function which drives landmark navigations and formal ways to assess the robustness of the whole approach can be introduced. In fact, the presence of a potential function around the goal is a sufficient condition for the application of the classical control theory based on Lyapunov functions. Tests performed with the *Nomad200* will conclude the paper (section 5).

2   Learning Landmarks

A landmark must be reliable for accomplishing a task and landmarks which appear to be appropriate for human beings are not necessarily appropriate for robots because of the completely different sensor apparatus and matching systems [31].

For example, the necessity of performing specific learning flights allows the insects to deal with objects which protrude from the background or which lie on a different plane than the background [13]. Attempts to understand in detail the significance of learning flights have been made only recently. Essentially, the flights are invariant in certain dynamic and geometric structures thus allowing the insects to artificially produce visual cues in specific areas of the eyes [35]. Perhaps, the main reason is that the precision for the homing mostly depends upon the proximity of chosen landmarks to the goal [12]. In fact, those flights need to be repeated whenever some changes in the goal position occur [23]. Therefore, it becomes crucial to understand whether or not a landmark is robust for the task accomplished by the agent.

Following the biological background and reconsidering the results presented in [31], one key point is that once the meaning of reliability has been established then the problem of selecting landmarks is automatically solved. Therefore, stating what is meant by *reliability of landmarks*, once given the specific sensor and matching apparatus, is a mandatory step.

2.1   Sensor and Matching Apparatus

The use of visual landmarks asks for real-time performances and this can lead to the use of specific hardware for their identification. The robot *Nomad200* (figure 1) that was used to accomplish the tests includes the *Fujitsu Tracking Card (TRV)* which performs real-time tracking of full color templates at a NTSC frame rate (30Hz). Basically, a template is a rectangular region of a frame which can be identified by two parameters m_x and m_y representing the sizes along X and Y axes respectively.

Fig. 1. The *Nomad200*

The card can simultaneously track many templates which have been previously stored in a video RAM. For each stored template the card performs a match in a sub-area of the actual video frame adopting the block matching method. This introduces the concept of *correlation* between the template and a sub-area of the actual video frame. The correlation measure is given by the sum of the absolute differences between the values of the pixels.

To track a template it is necessary to calculate the correlation between the template and a frame not only at one point on the frame but at a number of points within a *searching area*. The searching area is composed of 16×16 positions in the frame. The whole set of computed correlation measures is known by the term *correlation matrix*. Examples of correlation matrices are reported in figure 2. To perform the tracking, the matching system proposes as an output the coordinates of the position which represents the global minimum in the correlation matrix.

This approach strongly resembles the *region-based optical flow* techniques [2,24]. There, the flow is defined as the shift that yields the best fit between the image regions at different times.

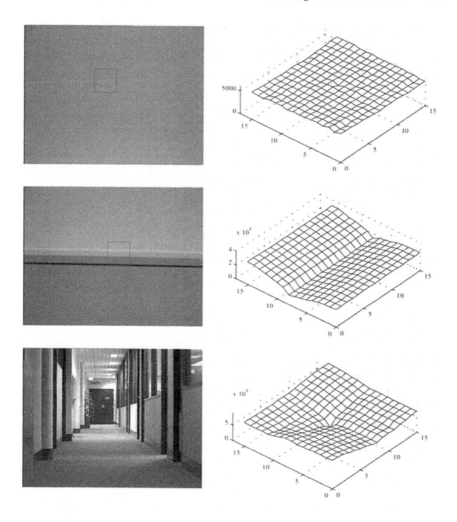

Fig. 2. Examples of correlation matrices which have been computed in the neighborhood of the templates (boxes in the pictures)

2.2 Choosing the Best Landmarks

Mori *et al.* have taken advantage of the correlation matrix to generate attention tokens from scenes [26]. The method they introduced is suitable to be used in the case of self-extraction of landmarks. They introduce a boolean measure to select a template:

$$1 - \frac{g}{g'} > \delta$$

where δ is a given threshold, g' is the (local) minimum value found in a circle (given a radius) around g (the *global minimum*) in the correlation matrix. Figure 3 exemplifies the method: g' is found in the gray area; g is the minimum value at the center of that gray area.

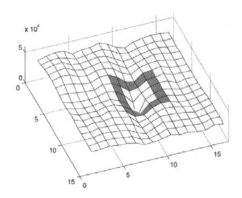

Fig. 3. The neighborhood used in the *Valley Method*

It is clear that this method (termed the *Valley Method* by the authors) is a crude approximation of a more sophisticated method that can explore the shape of the distortion matrix giving more information about the template and its neighborhood. However, the method works in real-time and for our applications this is crucial.

Relaxing the *valley method* the following can be obtained:

$$r = 1 - \frac{g}{g'} \tag{1}$$

where the value r is a measure on how the template is uniquely identifiable in its neighborhood: the greater r the more uniquely identifiable the template in its neighborhood. Therefore, by reliable landmarks we mean *templates which are uniquely identifiable.*

There are several degrees of freedom in searching for the best templates. Once having stated the definition of reliability, these degrees of freedom are represented by: the coordinates of the upper-left corner of the template being considered and the template size along X and Y. To avoid an extensive (and expensive) search in this 4-dimensional space we need to introduce some simplifications.

First of all we will consider only squared templates thus reducing the dimensions of the searching space by 1. This is a very substantial limitation, as instead of dealing with 64 possibilities we reduce this number to 8 because only eight different values along X and Y are handled by TRV. Some opportunities to deal with reliable templates are probably missed. In any case, dealing with squared templates doesn't undermine the whole method as we discovered performing the tests reported.

The second chance to speed up the search for the best template is to simplify the search for the co-ordinates of its upper left corner. In fact, regardless of the size of a template, we should select among 640 × 480 templates (this is the resolution of the NTSC signal processed by TRV), by comparing their distortion matrixes: even for a real-time tracking card like ours this computation is too big.

Therefore, for every template size m_{xy} (hereafter, this refers to $m_x = m_y$) we introduce the grid visible in figure 4. The grid has been designed following these principles:

- every cross represents the upper left corner of a possible template whose size is m_{xy}
- there must be enough room to compute the distortion matrix for each template

For these reasons we have different grids according to different m_{xy}. The first (top-left) cross is located in $(8 \times m_{xy}, 8 \times m_{xy})$ so as to leave enough room for computing one half of the distortion matrix. The same applies to the last (bottom-right) cross.

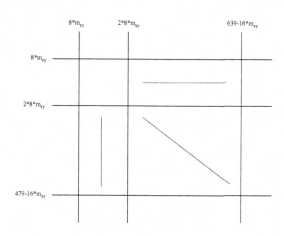

Fig. 4. The grid introduced to limit the amount of computation in searching for land-marks

As every cross represents the upper-left corner of a potential landmark and every template is $16 \times 16 \times m_{xy}^2$ in size, there is an overlap between two close templates, both along X and Y as shown in figure 5.

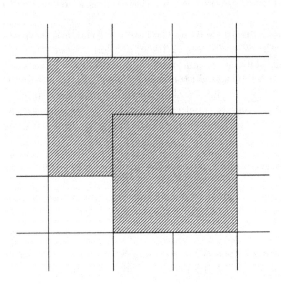

Fig. 5. Overlapping between neighboring templates

From a computational point of view the maximum number of selectable templates is $(640\ mod\ m_{xy}) \times (480\ mod\ m_{xy})$. Furthermore, a potential landmark is not independent of the neighboring ones (because of the overlapping). Once again, we should be aware that several combinations of templates are missed but considering the results of the performed tests this choice does not undermine the whole strategy.

Another limitation we will introduce concerns the freedom of size. Instead of choosing templates which give the opportunity to search for the best combination of size and upper-left corner, we keep the size fixed to a given value, for example $m_{xy} = 5$.

In conclusion, in order to select the best templates (so they can be referred to as *landmarks*) we maximize the following:

$$(o_x^*, o_y^*) = \arg \max_{\substack{(o_x, o_y) \in \text{grid} \\ m_{xy} \text{ fixed}}} r_l(o_x, o_y) \tag{2}$$

where $r_l(o_x, o_y)$ identifies the reliability factor for a landmark l whose upper left corner is located in (o_x, o_y). Therefore the position (o_x^*, o_y^*) with the highest r is returned. In order to assure that landmarks occupy different positions, previously chosen coordinates are not considered. Examples of landmarks chosen are reported in figure 6.

Obviously a landmark can have a different size and position from the original one. For example, when the robot is far from the goal place. Therefore, generally,

Fig. 6. Different choices of landmarks for different template sizes

the matching phase during the navigation task has to deal with the problem of guessing which is the actual size and position of a landmark. We can solve the problem using an extension of the classical block matching method implemented in TRV. If we suppose that in the set composed by the eight measures of distortion, the best value is represented by the minimum one (following the same principle that drives the block matching method), then we discover which is the *actual* best size. That is:

$$m_{xy}^* = \arg \min_{1 \leq m_{xy} \leq 8} D(o_x, o_y, m_{xy}) \tag{3}$$

where $D(o_x, o_y, m_{xy})$ represents the correlation value measured when the upper left corner of the template is located in (o_x, o_y) and its size is m_{xy}.

Once the best landmarks have been chosen from the static image of the goal, then they are stored in an internal video RAM to be used for successive tasks, the first being the *Turn Back and Look* phase.

2.3   The Turn Back and Look Phase

The landmarks which have been *statically* chosen will be used for navigation tasks. We found that it was necessary to *test* them in order to verify whether they represent good guides for navigation tasks.

TBL can help in verifying this [13,3] by testing whether during the motion the statically chosen landmarks still remain robustly identifiable.

Through a series of stereotype movements small perturbations (local lighting conditions, changes in camera heading, different perspectives and so on) can influence the reliability of the statically chosen landmarks. Referring to figure 7, the robot goes from the top to the bottom and the camera (arrows pointing to the top) is continuously pointing towards the goal.

These sorts of perturbations occur in typical robot journeys thus allowing us to state that the TBL phase represents a *testing framework* for landmarks. In other words, the robot tries to *learn* which landmarks can be suitably used in real navigation tasks by simulating the conditions the robot will encounter along the paths.

At the end of the TBL process only those landmarks that are visible and whose reliability r_l is above a given threshold are suitable to be used in navigation tasks.

The reliability factor r_l for landmark l is continuously computed during the TBL phase through the following:

$$r_l = \frac{\sum_{i=1}^{TBL} r_l^i}{TBL} \tag{4}$$

where TBL is the total number of steps exploited till that time, and r_l^i is the reliability of landmark l calculated at time i. In the tests, at the end of the phase, TBL usually consists of 400 steps (an internal counter) and it takes about 13 seconds to be performed. In figure 8 two pictures taken during a TBL phase exploited by the robot are shown.

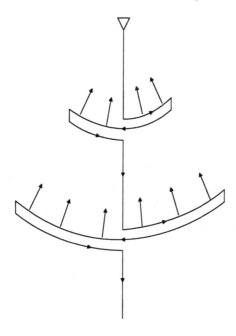

Fig. 7. The arcs performed by the robot to implement the TBL phase

3   Navigation from Landmarks

After reliable landmarks have been chosen then navigation information can be extracted from them [3,7]. The underlying biological principle is that a real movement is represented by an *attraction* force. It is produced by taking into account that the agent tries to restore the original position and size of the involved landmarks [9]. In fact, during the navigation task we are continuously computing the displacement vector (the difference between the actual position and the original one) and the actual size. Let d_l the difference between the original and the actual position of a landmark l and W_l a weight thus computed:

$$W_l = \begin{cases} \frac{M_{xy}}{m_{xy}} \text{ if } \frac{M_{xy}}{m_{xy}} > 1 \\ -\frac{m_{xy}}{M_{xy}} \text{ otherwise} \end{cases} \tag{5}$$

where M_{xy} is the original landmark size. The weight W_l takes into account if a landmark has a bigger size than the original one; in this case W_l is negative but still $|W_l| \geq 1$. The attraction given by a landmark l is:

$$v_l = d_l \cdot W_l \tag{6}$$

The data can be fused together by weighting them by introducing a sigmoid function $s(r_l)$ ranging from 0 to 1. The overall navigation vector can be thus

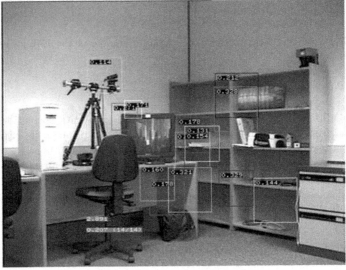

Fig. 8. Two pictures taken from a TBL phase: the numbers associated with each land-mark represent r_l in different times

calculated as:

$$V = [V_x \ V_y] = \frac{\sum_{l=1}^{L} v_l \cdot s(r_l)}{\sum_{l=1}^{L} s(r_l)} \tag{7}$$

where L is the number of landmarks chosen after TBL, r_l is the reliability value of landmark l and v_l is the attraction force *exerted* by landmark l. Lastly, V_x and V_y represent an estimation of the distance (along x and y axes of the environment) from the actual position to the goal position.

Figure 9 summarizes the situation where the picture represents a typical frame taken during a navigation test. In particular, the segment pictured in the circle at the bottom-center is proportional to the overall attraction exerted by the goal (see Eqn. 7). Above the circle the variance of that attraction is reported and under the circle the attraction vector is broken down into a module and an angle. In the circle on the right the single attraction exerted by each landmark is drawn. Each landmark has associated with it a number given by the value of the sigmoid function applied to its reliability measure. The arrows at the top-center of the figure represent the motion commands given to the robot. These translations are respectively the x and y component of the movement to be taken from this position to reach the goal.

Vector V represents the next movement with the module and direction relative to the actual robot position. The system dynamical model is therefore given by:

$$\begin{cases} x(k+1) = x(k) + V_x(x(k), y(k)) \\ y(k+1) = y(k) + V_y(x(k), y(k)) \end{cases} \tag{8}$$

where $x(k)$ and $y(k)$ represent the coordinates of the robot at step k; $V_x(x(k), y(k))$ and $V_y(x(k), y(k))$ the displacements computed at step k following Eqn. 7. These displacements are related to the position at step k given by $(x(k), y(k))$. Lastly, $x(k+1)$ and $y(k+1)$ represent the new positions the robot will move to. Clearly, an important equilibrium point (x^*, y^*) for the system is given by the coordinates of the goal position.

The navigation vector computed in Eqn. 7 can be considered as the overall attraction exerted by the goal position from that place.

4   The Guidance Principle

The system introduced by Eqn. 8 needs to be analysed in some detail. Basically, the dynamic system presented can be considered continuous-time with the following (omitting the vector notations):

$$\dot{x}(t) = V(x(t)) \tag{9}$$

where x represent the generic coordinates and an equilibrium point x^* is located at the goal position.

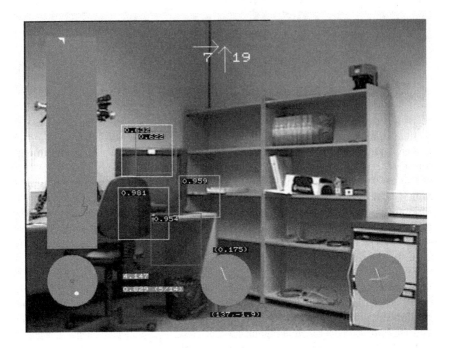

Fig. 9. A frame taken during a real navigation task. The segment pictured in the circle at the bottom-center is proportional to the overall attraction exercised by the goal (see Eqn. 7). Above the circle the variance of that attraction is reported and under the circle the attraction vector is broken down into a module and an angle. In the circle on the right the single attraction exercised by each landmark is drawn. Each landmark (box-shaped) has associated with it a number given by the value of the sigmoid function applied to its reliability measure: this is necessary to perform data fusion. The arrows at the top-center of the figure represent the motion commands given to the robot

Several important considerations for the stability of the system can be expressed focusing attention on its properties. In particular, when a dynamic system can be represented by $\dot{x} = f(x)$ with a fixed point x^*, and it is possible to find a *Lyapunov function*, i.e. a continuously differentiable, real-valued function $U(x)$ with the following properties [30]:

1. $U(x) > 0$ for all $x \neq x^*$ and $U(x^*) = 0$
2. $\dot{U}(x) < 0$ for all $x \neq x^*$ (all trajectories flow *downhill* toward x^*)

then x^* is globally stable: for all initial conditions $x(t) \to x^*$ as $t \to \infty$.

The system depicted in Eqn. 9 is of type $\dot{x} = f(x)$ but, unfortunately, there is no systematic way to construct Lyapunov functions [25].

A key point to solve the problem is represented by the *navigation vector field* the system produces [5,14,11,6,7]. An example of a navigation vector field is represented by figure 10. The goal (represented by a small circle) seems to be located within a basin of attraction.

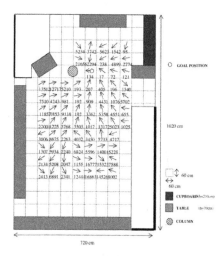

Fig. 10. An example of a navigation field: directions and modules (numbers)

In practice, the whole environment is subdivided into cells and for each of them the robot computes the navigation vector as given by Eqn. 7. The entire set of vectors is the navigation vector field.

Formally speaking, a vector field in two dimensions is a function that assigns to each point (x, y) of the xy-plane a two-dimensional vector $\boldsymbol{V}(x, y)$ usually represented by its two components:

$$\boldsymbol{V}(x, y) = [V_x(x, y) \; V_y(x, y)] \tag{10}$$

where $V_x(x, y)$ is the x-component and $V_y(x, y)$ is the y-component.

Particularly important is the *conservative* vector field. This is defined as one of which the integral computed on a closed path is zero, i.e. the vector field $\boldsymbol{V}(x, y)$ is conservative if and only if:

$$\oint_c \boldsymbol{V}(x, y) \circ d\boldsymbol{r} = 0 \tag{11}$$

for any closed path c contained in the field of $\boldsymbol{V}$; $d\boldsymbol{r}$ is the infinitesimal direction of motion. Such a field can always be represented as the gradient of a scalar function defined by:

$$U(x, y) = \int_{(0,0)}^{(x,y)} \boldsymbol{V}(X, Y) \circ d\boldsymbol{r} \tag{12}$$

where the path of integration is arbitrary. The scalar function U is referred to as the *potential* of the conservative force V in question. If the function $U(x,y)$ is known then the vector field can be determined from the relation:

$$V(x,y) = \nabla U(x,y) \tag{13}$$

All these considerations can be applied to our case. Assuming that the goal position is located at the minimum of the potential basin, Eqn. 13 is slightly modified with the following [21,19]:

$$V = [V_x \ V_y] = -\left[\frac{\partial U(x,y)}{\partial x} \ \frac{\partial U(x,y)}{\partial y}\right] \tag{14}$$

If the field is conservative, the scalar product introduced in Eqn. 12 can be further simplified by following a *particular* curve c [27]:

$$U(x,y) = -\int_{p_x}^{x} V_x(X, p_y)dX - \int_{p_y}^{y} V_y(x, Y)dY \tag{15}$$

where $U(x,y)$ is the potential function and the path of integration is along the horizontal line segment from the reference point (p_x, p_y) to the vertical line through (x,y) and then along the vertical line segment to (x,y). Every point can be referred to in terms of potential with the goal position. The problem of finding a Lyapunov function for Eqn. 8 can thus be solved numerically. If the goal place meets the two main Lyapunov criteria (see above) then the trajectories followed by the robot converge towards it.

The convergence to the goal is one of the main features related to the concept of robustness, Another key point is the repeatability of the experiments. Qualitatively speaking, the system converges to the goal but does it follow the same path when it departs from the same starting point? The analysis of this case involves a different property of a vector field: its *conservativeness*. This follows from a theoretical results. In particular, a non-conservative field is one in which the circuitation is non-null: the value of U computed by Eqn. 12 depends on the path followed. In other words, U is not determined entirely by the extreme points. Therefore the integration process can lead to an infinite set of results depending on the integrating path c. This means that diverse functions can drive the system dynamics thus producing different navigating paths. In addition, the convergence to the goal depends on the actual path chosen.

Therefore, a practical application of Eqn. 11 to state if the field is conservative needs to be found. To this extent, let us suppose that V_x and V_y are smooth and continuously differentiable and that the vector field is defined on a connected set. Under these hypotheses, a necessary and sufficient condition for the unique integration of the vector field is that the following relation (*Cauchy-Riemann*) holds:

$$\frac{\partial V_x(x,y)}{\partial y} = \frac{\partial V_y(x,y)}{\partial x} \tag{16}$$

In other terms:

$$\frac{\partial V_x}{\partial y} - \frac{\partial V_y}{\partial x} = 0 \tag{17}$$

The first member of Eqn. 17 can be a measure of the level of *conservativeness* of the vector field.

In conclusion, if the visual potential function has a basin of attraction with the minimum located at the goal position then the theory states that the system is intrinsically stable, at least starting navigating from within the convergent region of the environment.

5   Tests

As previously explained, tests have been performed both to measure how TBL affects the conservativeness of the navigation field and to calculate the region of convergence for the overall system.

The collection of the whole set of vectors (see Eqn. 10) is performed firstly placing the robot in a known position of the environment and then applying the method detailed in Eqn. 7. The iteration of the method over the whole environment and the collection of every displacement vector produces a vector field, as previously exemplified in figure 10. Each cell is approximately as big as the base of the robot.

Referring to Eqn. 17 the partial derivatives of V_x and V_y (respectively referred to as V_{xy} and V_{yx}) must be computed. Figure 12 plots the components V_x and V_y and their cross derivatives $\frac{\partial V_x}{\partial y}$ and $\frac{\partial V_y}{\partial x}$ taken by a real test.

The conservativeness of the field computed with a threshold set to 0 and landmarks sized 6 is shown in figure 11. Only small regions of the whole area roughly satisfy the constraint. A small threshold for TBL can dramatically change the situation. In figure 13 the amount of conservativeness for each point is plotted.

A key consideration is concerned with the scale along z: it is about one order of magnitude less than the one reported in figure 11. A trend toward a conservative field is thus becoming evident.

The situation obtained with a threshold of TBL set to 0.2 has been reported in figure 14. A large area of the environment has a measure of conservativeness that roughly equals 0.

Similar considerations can be expressed dealing with a different landmark size. For example, figure 15 shows the case where the TBL threshold is 0.2 and landmarks have a size of 4. The *template* of the graph is the same as before. Therefore, with a good choice of the threshold the field becomes conservative regardless of the size of the landmarks.

5.1   Computation of the Visual Potential Field

The computation of the visual potential field must be performed only on those areas of the environment which are conservative. From the results of the tests only two cases can be considered: when TBL threshold is set at 0.2 and landmarks have a size of 6 or 4. Starting with the former and following the method detailed in the previous section, the visual potential function is shown in figure 16.

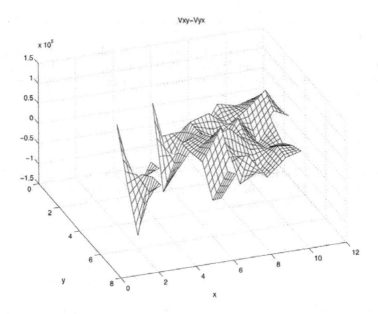

Fig. 11. Conservativeness of a vector field computed with a TBL threshold of 0 and landmarks sized 6

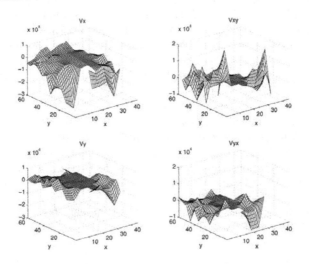

Fig. 12. An example of V_x and V_y and their cross derivatives. In particular, V_{xy} refers to $\frac{\partial V_x}{\partial y}$ and V_{yx} refers to $\frac{\partial V_y}{\partial x}$

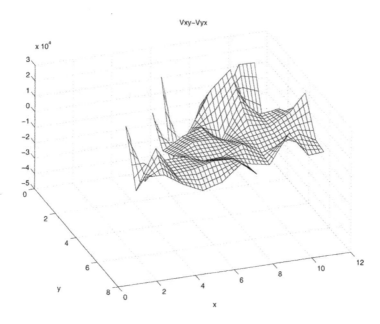

Fig. 13. Conservativeness of a vector field computed with a TBL threshold of 0.1 and landmarks sized 6

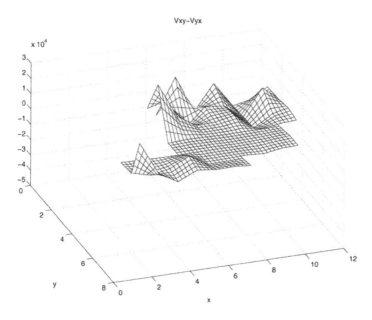

Fig. 14. Conservativeness of a vector field computed with a TBL threshold of 0.2 and landmarks sized 6

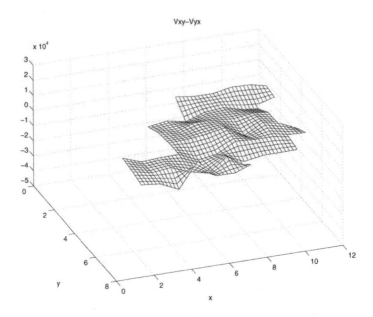

Fig. 15. Conservativeness of a vector field computed with a TBL threshold of 0.2 and landmarks sized 4

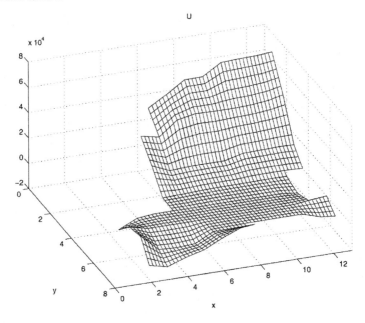

Fig. 16. Potential function computed with a TBL threshold set to 0.2 and landmarks sized 6

The goal position is located in $(4,5)$ and it represents the *reference point* for Eqn. 15. The shape of the potential function tends to produce a minimum around the goal (see figure 17).

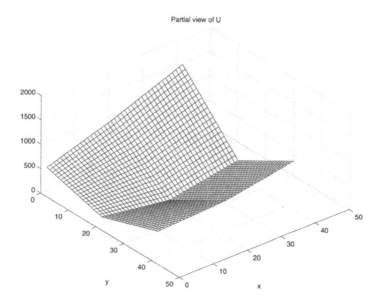

Fig. 17. Potential function computed with a TBL threshold set to 0.2 and landmarks sized 6 magnified 30 times around the goal

In addition, roughly speaking, the basin of attraction of the goal is the whole environment (where the tests were performed), i.e. apart from some cases, all the starting points lead to the goal. To this extent, consider the differences in the potential field showed in figure 19 where the TBL threshold is 0.2 but the landmark size is 4.

There are two important differences: the first concerns the basin of attraction and the second is concerned with the depth of the minimum in the goal position. The basin of attraction determines how far a goal position can be *felt*. In other words, if the robot starts navigating within the basin of attraction then it reaches the goal position. Outside the basin the robot could lead to other (false) goals. The visual potential function reported in figure 16 possesses a larger basin of attraction than the one reported in figure 19 which influences the robot only when the robot is close to the goal.

Intriguingly enough, this has strong analogies with the biological results reported in [10]. The authors discussed the area of attraction of the goal (namely, *the catchment area*) considering the size of landmarks surrounding it. Large landmarks determine larger catchment areas than smaller landmarks. Furthermore,

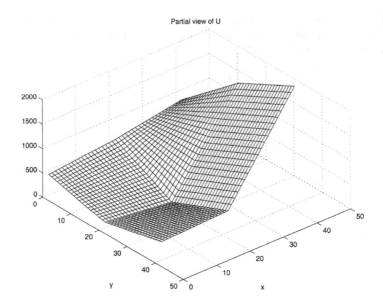

Fig. 18. Potential function computed with a TBL threshold set to 0.2 and landmarks sized 4 magnified 30 times around the goal

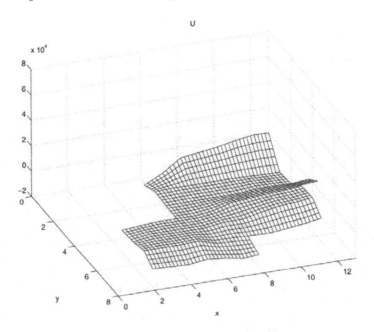

Fig. 19. Potential function concerning the case with a TBL threshold set to 0.2 and landmarks sized 4

larger landmarks determine a coarse approach to the goal whereas smaller land-marks allow the insects to precisely pinpointing the goal position.

To this extent, the visual potential function produced with landmarks sized 4 has a deeper minimum at the goal than the potential obtained with a size of 6 (see figure 18 compared to 17).

Lastly, a comparison between the following two figures can witness the amount of robustness gained by the system when learning involves big TBL thresholds. The first test conducted and reported in figure 20, doesn't consider any TBL selection phase.

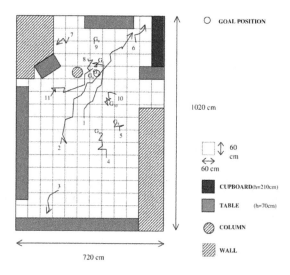

Fig. 20. Experiments conducted without any TBL phase with landmarks sized 6

The landmarks used were sized was 6 (that is $m_{xy} = 6$) and there were 14 of them. Every navigation trial has been numbered from 1 to 11 and every individual ending point has been reported with the notation G_i where i is the starting point of navigation i. Sometimes, some navigation trials has no end within the environment so its path has been labeled with an arrow at the point where it finishes.

On the other hand, we also need to consider navigation trials accomplished *after* a TBL selection (threshold 0.10) as reported by figure 21. In this situation, 14 landmarks with a size of $m_{xy} = 6$ were originally selected. After a TBL phase that number was reduced to 8. Now the navigation paths are smoother compared to those previously reported. Nevertheless, there are some navigations (7, 3 and 4) whose ending points do not lie in the right position. Those starting points (at least for trials number 3 and 4) can be easily trapped by other minima other than the main goal.

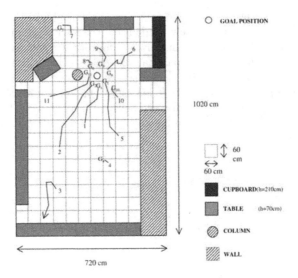

Fig. 21. Experiments conducted with a TBL phase and landmark sized 6

5.2   Issues on the Visual Potential as a Lyapunov Function

From the potential function previously plotted it can be easily understood why the system gets sometimes trapped into false goals or what can be the region of convergence for the main goal position.

This implicitly states that the system has no overall stability on the whole environment. Therefore, the visual potential function itself cannot be considered Lyapunov compliant unless reducing its domain of application to a region around the goal position, starting from which the system converges (see e.g. figure 17).

6   Related Work

To date, biologically-based methods have not been widely addressed in robotics. Very recently, however, in the survey paper by Trullier *et al.* [32] and in books by Srinivasan and Venkatesh [29] and Aloimonos [1] some interesting adaptations of methods directly inspired by animals and insects have emerged. This, however, represents more a niche development than a classical approach considering that even recent reviews that can be defined *classical*, e.g. [21,17,8], do not address any biologically-inspired methods.

Basically, a mile-stone for visual guidance mechanisms is represented by [9], where a theoretical computational model to explain the Bee visual guidance behavior is introduced. Some interesting extensions to the model introduced in [9] are addressed in a recent work [34]. The main contribution concerns the use of a pyramidal structure for landmark matching between current and snapshot images that the author considers as uni-dimensional and gray-level.

An approach to grey-level bi-dimensional images is addressed by Hong *et al.* [16]. The authors use an omnidirectional device (a hemisphere) for extracting the whole panorama. The robot computes a one-dimensional circular location signature (an array of 256 elements ranging from 0 to 255) by sampling the hemispherical image along a precise radius at angular intervals. Characteristic points of this signature are then found by investigating the brightness profile. A matching phase is then performed by comparing the signatures (the one taken at the goal and the one computed considering the actual image) by an affine model. The affine factors are estimated at a preliminary stage using the signatures extracted.

The use of affine models to compute the final navigation vector has been addressed by Salas *et al.* [28]. Here, the authors try to recognize and locate artificial landmarks scattered across a workspace using algebraic invariants for landmark recognition and matching. Those invariants do not change under general affine transformations. Once the shape has been recognized the inverse problem consists of finding the parameters of the geometrical transformation. The parameters are directly mapped to real robot displacements.

Recently, a log-polar transformation for recognizing the most suitable landmarks in the environment has been considered by Gaussier *et al.* [15]. The landmarks are learnt during a preliminary exploration phase of the robot around the goal. They are represented by a log-polar transformation of a quasi-panoramic image (see the work of Joulain *et al.* [18] for a detailed explanation of the methods used to extract salient landmarks from the environment). The learning module is represented by a neural network. Later during the navigation phase the system gives as an input to the neural network the actual (transformed) image and its level of activation gives enough informations on the recognition of previously learnt landmarks and their position values in the environment.

In Bianco *et al.* [4] an extension of the models presented in [9,16,34,28] that deals with full resolution images has been considered. In order to estimate the robot movement, a comparison is made between the actual image and the one representing the goal place. All the possible changes due to robot movements can be described by a simplified version of a classical affine model.

A completely different approach has been addressed by Lambrinos *et al.* [20]. In this work, in particular, a new model called the *Average Landmark Vector (ALV)* is introduced: it explains the insect's navigation capabilities without considering any snapshot or, of course, any matching phase. The authors think that the only piece of information needed is the direction of the average landmark vector acquired by a summation of unit vectors pointing to landmarks. The target direction is calculated in a specific position as is the difference between the AL vector and the one at the current position. The ALV model shows very robust behavior in complex realistic environments where several landmarks are present.

7 Concluding Remarks

This paper has presented the learning system of a biologically-inspired navigation method based on natural visual landmarks. The visual learning phase (TBL) affects the conservativeness of the navigation vector field thus allowing us to explain landmark navigation in terms of a potential field.

Conversely, the computation of the conservativeness of a navigation field can assess the reliability of the landmarks chosen and, therefore, measure the quality of the learning phase.

Lastly, the presence of a potential function around the goal allows us to apply classical control theories to assess the robustness of the overall navigation system. The idea of applying methods from vector analysis to navigation problems allows us to evaluate the performance of different models and can represent an important step for topological navigations.

References

1. Y. Aloimonos. *Visual Navigation from Biological Systems to Unmanned Ground Vehicles*. Lawrence Erlbaum Associates, Publishers, Mahwah, New Jersey, 1997.
2. P. Anandan. A computational framework and an algorithm for the measurement of visual motion. *International Journal of Computer Vision*, (2):283–310, 1989.
3. G. Bianco. *Biologically-inspired visual landmark learning and navigation for mobile robots*. PhD thesis, Department of Engineering for Automation, University of Brescia (Italy), December 1998.
4. G. Bianco, R. Cassinis, A. Rizzi, N. Adami, and P. Mosna. A bee-inspired robot visual homing method. In *Proceedings of the Second Euromicro Workshop on Advanced Mobile Robots (EUROBOT '97)*, pages 141–146, Brescia (Italy), October 22-24 1997.
5. G. Bianco, R. Cassinis, A. Rizzi, and S. Scipioni. A proposal for a bee-inspired visual robot navigation. In *Proceedings of the First Workshop on Teleoperation and Robotics Applications in Science and Arts*, pages 123–130, Linz (Austria), June 6 1997.
6. G. Bianco, A. Rizzi, R. Cassinis, and N. Adami. Guidance principle and robustness issues for a biologically-inspired visual homing. In *Proceedings of the Third Workshop on Advanced Mobile Robots (EUROBOT '99)*, Zurich (Switzerland), September 1999.
7. G. Bianco and A. Zelinsky. Biologically-inspired visual landmark learning for mobile robots. In *Proceedings of the IEEE/RSJ International Conference on Intelligent Robots and Systems*, Kyongju (Korea), October 17-21 1999.
8. J. Borenstein, H.R. Everett, and L. Feng. *Where am I? Sensors and Methods for Mobile Robot Positioning*. The University of Michigan, April 1996.
9. B.A. Cartwright and T.S. Collett. Landmark learning in bees. *Journal of Comparative Physiology*, A(151):521–543, 1983.
10. B.A. Cartwright and T.S. Collett. Landmark maps for honeybees. *Biological Cybernetics*, (57):85–93, 1987.
11. R. Cassinis, A. Rizzi, G. Bianco, N. Adami, and P. Mosna. A biologically-inspired visual homing method for robots. In *Proceedings of seventh workshop of AI*IA on Cybernetic and Machine Learning*, Ferrara (Italy), April 1998.

12. K. Cheng, T.S. Collett, A. Pickhard, and R. Wehner. The use of visual landmarks by honeybees: Bees weight landmarks according to their distance from the goal. *Journal of Comparative Physiology*, A(161):469–475, 1987.

13. T.S. Collett and J. Zeil. Flights of learning. *Journal of the American Psychologycal Society*, pages 149–155, 1996.

14. P. Gaussier, C. Joulain, and J. Banquet. Motivated animat navigation: a visually guided approach. In *Simulation of Adaptive Behavior*, Zurich (Switzerland), August 1998.

15. P. Gaussier, S. Lepetre, C. Joulain, A. Revel, M. Quoy, and J. Banquet. Animal and robot learning: experiments and models about visual navigation. July 1998.

16. J. Hong, X. Tan, B. Pinette, R. Weiss, and E.M. Riseman. Image-based homing. In *Proceeding of the IEEE International Conference of Robotics and Automation*, pages 620–625, Sacramento, April 1991.

17. Y.K. Hwang and N. Ahuja. Gross motion planning - a survey. *ACM Computing Surveys*, 24:219–291, September 1992.

18. C. Joulain, P. Gaussier, A. Revel, and B. Gas. Learning to build visual categories from perception-action associations. In *Proceedings of Intelligent Robots and Systems (IROS)*, pages 857–864, Grenoble (France), 1997.

19. O. Khatib. Real-time obstacle avoidance for manipulators and mobile robots. *Int. Journal of Robotics Research*, 4(1):90–98, 1986.

20. D. Lambrinos, R. Moeller, R. Pfeifer, and R. Wehner. Landmark navigation without snapshots: the average landmark model. In *Proceedings of Neurobiol. Conference*, Goettingen, 1998.

21. J.C. Latombe. *Robot motion planning.* Kluwer Academic Publisher, Boston/Dodrecht/London, 1991.

22. M. Lehrer. Why do bees turn back and look? *Journal of Comparative Physiology*, A(172):549–563, 1993.

23. M. Lehrer. Honeybees' visual spatial orientation at the feeding site. In M. Lehrer, editor, *Orientation and communications in arthropods*, pages 115–144. Birkhauser verlag, Basel/Switzerland, 1997.

24. J.J. Little and A. Verri. Analysis of differential and matching methods for optical flow. In *Proceedings of the IEEE workshop on visual motion*, pages 173–180, Irvine, CA, 1989.

25. D.G. Luenberger. *Introduction to dynamic systems - theory, models, and applications.* John Wiley and Sons, New York Chichester Brisbane Toronto, 1979.

26. T. Mori, Y. Matsumoto, T. Shibata, M. Inaba, and H. Inoue. Trackable attention point generation based on classification of correlation value distribution. In *JSME Annual Conference on Robotics and Mechatronics (ROBOMEC 95)*, pages 1076–1079, Kavasaki (Japan), 1995.

27. C.C. Ross. *Differential Equations An Introduction with Mathematica.* Springer-Verlag, New York Berlin Heidelberg London Paris Tokyo Hong Kong Barcelona Budapest, 1995.

28. J. Salas and J.L. Gordillo. Robot location using vision to recognize artificial landmarks. In *SPIE*, volume 2354, pages 170–180, 1995.

29. M.V. Srinivasan and S. Venkatesh. *From Living Eyes to Seeing Machine.* Oxford University Press, Oxford/New York/Tokyo, 1997.

30. S.H. Strogatz. *Nonlinear dynamics and chaos.* Addison-Wesley Publishing Company, New York, 1994.

31. S. Thrun. A bayesian approach to landmark discovery and active perception in mobile robot navigation. Technical report, School of Computer Science Carnegie Mellon University, 1996.

32. O. Trullier, S.I. Wiener, A. Berthoz, and J.A. Meyer. Biologically based artificial navigation systems: Review and prospects. *Progress in Neurobiology*, 51:483–544, 1997.

33. R. Wehner. Arthropods. In F. Papi, editor, *Animal Homing*, pages 45–144. Chapman and Hall, London, 1992.

34. T. Wittman. Insect navigation: Models and simulations. Technical report, Zentrum fur Kognitionswissenschaften, University of Bremen, 1995.

35. J. Zeil, A. Kelber, and R. Voss. Structure and function of learning flights in bees and wasps. *Journal of Experimental Biology*, 199:245–252, 1996.

Author Index

Lecture Notes in Artificial Intelligence (LNAI)

Lecture Notes in Computer Science